T0132520

Membranes to Molecular Machines

Synthesis

A series in the history of chemistry, broadly construed, edited by Carin Berkowitz, Angela N. H. Creager, John E. Lesch, Lawrence M. Principe, Alan Rocke, and E. C. Spary, in partnership with the Science History Institute

Membranes to Molecular Machines:
Active Matter and the Remaking of Life

Mathias Grote

The University of Chicago Press :: Chicago and London

The University of Chicago Press, Chicago 60637
The University of Chicago Press, Ltd., London
© 2019 by The University of Chicago
Published 2019
Printed in the United States of America

28 27 26 25 24 23 22 21 20 19   1 2 3 4 5

ISBN-13: 978-0-226-62515-7 (cloth)
ISBN-13: 978-0-226-62529-4 (e-book)
DOI: https://doi.org/10.7208/chicago/9780226625294.001.0001

Library of Congress Cataloging-in-Publication Data
Names: Grote, Mathias, author.
Title: Membranes to molecular machines : active matter and the remaking of life / Mathias Grote.
Description: Chicago : The University of Chicago Press, 2019. | Includes bibliographical references and index.
Identifiers: LCCN 2018044853 | ISBN 9780226625157 (cloth : alk. paper) | ISBN 9780226625294 (e-book)
Subjects: LCSH: Molecular biology—Research—History. | Membranes (Biology)—Research—History. | Biotechnology—Research—History.
Classification: LCC QH506 .G768 2019 | DDC 572.8—dc23
LC record available at https://lccn.loc.gov/2018044853

♾ This paper meets the requirements of ANSI/NISO Z39.48-1992 (Permanence of Paper).

To my parents, who taught me to look at small things

D'ALEMBERT: [. . .] Si c'est une qualité générale de la matière, il faut que la pierre sente.

DIDEROT: Pourquoi non?

D'ALEMBERT: Cela est dur à croire.

DIDEROT: Oui, pour celui qui la coupe, la taille, la broue et qui ne l'entend pas crier.

Denis Diderot, *Entretien entre Diderot et d'Alembert*, 1769.

Contents

A gallery of color plates follows page 128.

Preface

Why membranes, why molecular machines, why me? It may come as little surprise that someone embarking on historically uncharted territory such as this has previous experience with the subject. Membranes and proteins had been literally in my hands: From 2004 to 2008, I worked as a PhD student in a molecular biological laboratory of Humboldt University of Berlin, growing bacteria in large culture flasks, extracting their membrane fractions by centrifugation, and purifying proteins from them that were known to perform transport processes across the cellular membranes. My goal was to characterize the structure and dynamics of one such protein experimentally. Through years of sometimes pretty tedious lab work, the group that I was part of described bits and pieces of the process, or mechanism, by which this protein was able to push its freight across the membrane. Thus, the objects of this book, and their representations, similar to the ones depicted in plates 1 and 2, had been part and parcel of my daily work.

My itinerant scientific socialization—before pipetting, I had studied philosophy and turned to the history of science after my PhD—has certainly colored my take on the topic of this book, not least through the usage of terminology from different disciplines. You may have wondered reading the last paragraph about what membranes and proteins exactly are, and you may not want to switch to a screen to check. I have attempted to remedy this somewhat necessary evil of work on contemporary science through a short glossary of key terms at the end of this book. More importantly, my time working with membranes and proteins had put a topic on the map for me that hardly existed historically or philosophically.[1] Whereas heredity and the gene had received plenty of attention by humanities scholars, especially in the wake of the Human Genome Project around 2000, and while midcentury biochemistry had been scrutinized by a previous generation of historians of science, membranes and proteins remained a desideratum of study and only a few studies had scratched the surface of this subject.[2]

As someone with one foot still in the lab, the neglect of these topics looked utterly strange in light of the huge impact this research had made on contemporary science as well as on biotechnologies and medicine, but even more so since membranes seemed to reveal the life sciences in a different light. Sometimes, or so it seemed to me, impressive amounts of ink had been spilled to refute narratives that placed molecular genetics at the center of biology in the twentieth century; however, very few scholars dared to grab the plethora of fascinating histories beyond the gene. As membrane research was based on different conceptual and technological premises, and as it engendered a different picture of bodies, cells, and life, looking at it more closely would help to correct such historiographical artifacts.

Here are some new, membrane-based questions that went through my beginner's mind in the history of science, many still revealing my scientific as much as my philosophical past: What does it imply for our understanding of life, and for the capacity to act on it in the early twenty-first century, if we can take cells apart and put their molecular components back together to perform physiological processes in a test tube—a problem that has become obvious in recent discussion of synthetic biology, but that has been pertinent to membrane and cell biology at least since the interwar period? What I held in my test tubes were greasy lumps of active protein and lipid matter, supposedly standing in for the delicate membrane films that surround the living cell, and the molecular machinery sitting therein—what did this tell me about the materiality of life in contemporary laboratory science? Or, how could the concept of the membrane "transporter" I was tackling, an example of a machine-like molecule, be related to the models and narratives of molecular mechanisms that I had discussed in my thesis: Was this protein really some sort of mechanically functioning machine, or was this just a fancy way of talking about enzymes and biochemistry? Or, even more broadly, what did it imply for the concept of life, and many other central problems of biology, to look away for a moment from the problem of heredity and the informational language of midcentury molecular biology, and to look at protein molecular machinery as an instance of active matter—the latter being a topic at the crossroads of the history of the life science, chemistry, and nanotechnologies that was just gaining momentum while I was writing this book?

In order to address some of these huge questions historically, I set out on a case study on a well-researched molecular machine, in fact a "pump" that I knew from textbooks and lectures—more on that in the Introduction. Doing a close-up, or so I hoped, would also allow the much debated problem of molecular mechanisms in the philosophy of science to be viewed from another angle. When comparing the "ready-made," extremely simplified sketches of exemplary molecular mechanisms with my experiences from the lab, these seemed to somehow miss a central point: Even if experimentation was discussed here, the fact that scientists significantly impacted on organisms, cells, etc. in order to spell out these mechanisms seemed underappreciated in philosophical analyses. In how far had recent research on molecules and mechanisms transformed the materiality of life? Concisely, is the stuff that life is made of still the same that it was before these sciences set out to take cells apart and put them back together?

But there are also historiographical reasons to scrutinize the surge of membranes and molecular machines, some of which began to dawn on me only in the course of my study: In what follows, I will describe the work of a generation of influential protagonists from the 1970s to 1990s, who had been shaping a novel molecular biology in these years, and who were, at the time I was beginning this project, leaving their posts. In early 2009, I traveled from Berlin to Munich and met with biochemist Dieter Oesterhelt, director at the Max Planck Institute (MPI) of Biochemistry since 1979. He was an enthusiastic supporter of my project from the beginning, and soon we went through his notebooks and documents, in the midst of a still running laboratory. A few scanning sprees made significant parts of his papers available to me. A year later, while I was working at the University of Exeter in England, I profited from a second, similarly fortunate encounter: Richard Henderson, structural biologist and director of Cambridge's famous Medical Research Council Laboratory of Molecular Biology (LMB) from 1996 to 2006. Shortly after our first meeting, I was sitting in his attic, leafing through the printouts and photos from his electron microscopic studies of the 1970s. And as I am writing this preface seven years later, Richard Henderson has just shared the Nobel prize in chemistry for exactly this work. More-over, through interviews and conversations, looking at photos, confer-ence programs, and molecular models during my visits to the labs in Cambridge, Munich, Irvine, San Francisco, and Santa Cruz, I became immersed in the community of membranologists that had been flourish-ing when I was learning to walk and talk. Insofar, this book may also be read as a preliminary first insight into a generation of influential sci-entists of the near past, in a field and in institutions of science many of which have not garnered proper attention. Moreover, although most of the protagonists are alive and well, this past looks quite distant in many respects—certainly regarding the technological possibilities of research, but also the larger ramifications of the life sciences. Whereas many of my protagonists had very basic-looking projects on their hands, the field is nowadays strongly influenced by envisaged usages of "molecular ma-chinery" in nanotechnologies or the neurosciences. For an example, one could just mention optogenetics, a field in between membrane research, recombinant DNA, and neurobiology that attempts to influence nerve activity by engineering the molecular machinery the emergence of which will be discussed in this book. Insofar, this is also the history of a con-stellation still in flux, and one with many open questions about scientists and their objects.

Acknowledgments

In 2007, I was mostly pipetting. Maybe to get away from the lab for a moment, I signed up as a visitor for a workshop on the "Cultural His-tory of Heredity" at the University of Exeter, UK. Shortly before the meeting, I was offered a vacant room in the university dorm, as one of the participants had cancelled. So, quite unexpectedly, I was lucky enough to share English breakfasts with historians of the life sciences. As this was my introduction to a field I knew before only from books and classes, my first thanks go the two room brokers, Cheryl Sutton and Staf-fan Müller-Wille, both at Egenis at the University of Exeter, UK.

Only shortly after, Maureen O'Malley, John Dupré, and Hans-Jörg Rheinberger made it possible for me to embark on the project that led to this book. I would like to thank them for the exciting and fruitful years I spent as a postdoc at the MPI for the History of Science, Berlin (2009), and at Egenis (2010), but most of all for their confidence in taking a dis-ciplinary border crosser on board. Back in Berlin, the project was contin-ued, and in fact found much of its present shape, at the chair of Friedrich Steinle at the Technische Universität Berlin, whom I would like to thank for making me part of his newly forming group, and for supporting my grant application to the German Research Foundation (DFG), whose well-measured funding program of the "Eigene Stelle" allowed me to set up a tailor-made one-man project. One of the highlights of what appears in retrospect as a time of abundant freedom of research from 2011 to 2014 were my visits to the Centre Cavaillès at the Ecole Normale Supérieure in Paris, and the profound conversations I had with my host, Michel Morange. Another was my encounter with Angela N. H. Creager (Princeton), just as Michel another expert of all things molecular biological, and a kindred spirit sharing my enthusiasm for biochemistry and proteins in the 1970s and 1980s. A fellowship by the Chemical Heritage Foundation in Philadelphia (now the Science History Institute) in 2015 allowed me to "get my stuff together" and to conceive of this book in its present form—Carin Berkowitz and Carsten Reinhardt, among many others, are thanked for this opportunity and the great atmosphere at what was then CHF, which also brought me into contact with one of the few other membrane fans in the history of the life sciences, Daniel Liu. Finally, Anke te Heesen supported me with candid firmness in finishing this manuscript while I was actually doing many other things at Humboldt University. The spirit and the folks of Ringenwalde, where much of this final work took place, made these days and weeks sweet. And there are so many more historians and philosophers of science who inspired me or discussed this topic with me: Jenny Bangham, Soraya de Chadarevian, Christian Joas, Ursula Klein, Karin Krauthausen, Sabina Leonelli, Sacha Loeve, Robert Meunier, Kärin Nickelsen, Laura Otis, Christian Reiß, Max Stadler, Jim Strick . . .

Writing the history of recent science would not have been possible without insight from scientists. Therefore, I thank biophysicists Roberto Bogomolni (UC Santa Cruz) and Janos Lanyi (UC Irvine), Peter Hegemann (HU Berlin), Norbert Hampp (University of Marburg), and Hartmut Michel (MPI for Biophysics, Frankfurt), but most of all Richard Henderson (LMB, Cambridge, UK) and Dieter Oesterhelt (MPI of Biochemistry, Martinsried) for their attention, their time, and the stimulating conversations. Special thanks go to Richard Henderson and Dieter Oesterhelt for providing me access to their research papers.

Anke te Heesen, Daniel Liu, Cyrus Mody, Hanna Worliczek, and two anonymous reviewers are thanked for their comments on parts or prior versions of this manuscript and their advice on how to make things explicit. Vincent Dold, Laura Haßler, Konstantin Kiprijanov, Carla Seemann, Róisin Tangney, and Lotte Thaa have helped me at various stages of research or manuscript preparation. Thanks also to the Synthesis series of the Science History Institute, Philadelphia, for taking me on board, and especially to Karen Merikangas Darling and colleagues at the University of Chicago Press for encouraging and stimulating discussions on how to turn an academic manuscript into a book.

Research on this book was funded by the German Research Foundation DFG (grant GR 3835–1/2 to M.G.). The Introduction and Chapter 2 draw on some material prior versions of which have been published in Grote 2013a and Grote 2014. The Syndics of Cambridge University Library are acknowledged for permission to quote from Peter Mitchell's papers.

Introduction: The Molecular-Mechanical Vision of Life

Enzymes are awesome machines with a level of complexity that suits me.

Arthur Kornberg, 1989, p. 299

Organisms, and so ourselves, composed of tiny machines may be an unfamiliar thought. Are we not made up of flesh and blood, and fibers, vessels and bone? Yet, the idea that our cells are composed of molecular machinery made of specific, active matter responsible for our bodily processes from digestion to movement to perception may be closer to our everyday lives than we think. In the coming pages, I will introduce some basic scientific premises and concepts of what I consider as a powerful, materialistic vision of life of our days, before setting out to discuss the epistemological and the historical basis of this vision, the study of which forms the topic of this book.

My little tour through the contemporary life sciences starts almost as close as it can get to us: in our stomach. Imagine suffering from heartburn. In the industrialized world, the most likely advice from a physician—apart from changes in lifestyle or diet—would be to reduce the excess acidity in your stomach, using a pill containing a substance called a "proton pump inhibitor." In fact, these pharmaceuticals count amongst the biggest sellers worldwide. Proton pump inhibitors, or so Wikipedia will tell a curious patient, modify the action of tiny "pumps," proteins that sit in our cells' membranes, that is, in the thin lipid films, which form the boundary in between cell and environment. The pharmaceutical will thus "block" the action of these specific proteins, in this case those sitting in the cell membranes of the gastric mucosa. After ingestion of a proton pump inhibitor pill, gastric mucosa cells excrete fewer protons into the stomach, leading to less acid production. In other words, the substance alters the mode of operation of our body's "molecular machinery," thus modifying cellular physiology. Problem solved? It is exactly this way of thinking and acting, or so I believe, that makes the molecular-mechanical vision so appealing to researchers, the pharmaceutical industry, and the health care market as well as patients or other individuals alike. If we take into account that many drugs used to treat psychic phenomena from anxiety to insomnia, schizophrenia, or depression, described as working in a similar fashion, we can estimate the reach and the implications of this vision. The controversial pharmaceuticals called SSRIs (or "selective serotonin reuptake inhibitors," better known under such brand names as Prozac), for example, are thought to act as "molecular wedges," interfering with "transporter" proteins, which sit in nerve cell membranes and facilitate the re-uptake of the neurotransmitter serotonin from the synaptic cleft between the cells. What is more, the molecular-mechanical vision is not at all restricted to pharmaceuticals: The plant poison curare, for instance, which is used to poison hunting arrows as well as in surgery to relax muscles, or so we are told in today's textbooks, achieves its stunning effect by "blocking" the muscle's acetylcholine-receptors, and in fact all sorts of physiological processes are explained by specific protein machinery moving, being blocked, pushing something, reacting with certain substances, etc.[3] One caveat beforehand: In what follows, my aim is neither to legitimize certain drugs, nor a specific way of conceiving of and acting upon organisms that is widespread today. My aim is to show how science and technology got to this point in the last half-century or so, and I will do so by following where concepts of molecular machinery have appeared, how they have materialized, and how they were put into practice in the laboratory. In brief, when I discuss how a "vision" of how to understand and act upon life became so powerful, I claim neither that molecular-mechanical models of physiological processes are epistemically adequate or productive at all times, nor that they are the only way of conceiving of life. And, needless to say, I do not want to advocate that taking biomedical pills is always the best way to cure an ailment—eating less greasy food may do better in the case of a heartburn, which illustrates the social aspect of the seemingly esoteric scientific topic of molecular machinery.[4] However, it remains uncontroversial that the molecular-mechanical vision is extremely influential in today's life sciences, biotechnology, and medical practice. As this book will show, it has become a "winner's perspective" for how to conceive of and act upon life, and it is in this sense that it may be comparable to the midcentury rise of the "molecular vision" of life described by Lily Kay, when it comes to epistemic dominance, to dissemination and popularization, as well as, presumably, to funding or institutional support in the past decades.[5]

But back to science for a moment: The mentioned drugs are complex organic molecules binding to their target "machines" on the basis of chemical specificity, similar to how an antibody recognizes a bacterium, or how an enzyme recognizes its specific substrate, namely according to a so-called lock-and-key model as described by organic chemist Emil Fischer around the turn of the twentieth century. Yet, the pivotal question of how contemporary science understands the target machinery requires further explanation. *The Machinery of Life* is a richly illustrated atlas of life at the microscale, which was first published by illustrator and molecular biologist David Goodsell in the early 1990s. This was not at all the first book to present a popularizing image of life's molecules—under the title *Mr. Tompkins inside Himself*, physicist George Gamow and biologist Martinas Yčas, for example, had published the "adventures in the new biology" as a voyage into the molecular body already in 1968. Yet, Goodsell's book allows us to get an impression of the contemporary molecular cosmos in which proteins, DNA, lipids, and other substances of life act and interact. Similar to the famous 1977 film *Powers of Ten* by Charles and Ray Eames, which zooms in from the hand of a picnicker at Chicago's lakeside to the surface of the skin, into the cells, and finally into the atomic makeup of the molecules composing the cells, Goodsell takes the reader on a trip through our interior micro-universe:

> The human body is a living, breathing example of the power of nanotechnology. Almost everything happens at the atomic level. Individual molecules are captured and sorted, and individual atoms in these molecules are shuffled from place to place, building entirely new molecules. Individual photons of light are captured and used to direct the motion of individual electrons through electrical circuits. Molecules are packaged and transported expertly over distances of a few nanometers. Tiny molecular machines [. . .] orchestrate all of these nanoscale processes of life. Like the machines of our modern world, these machines are built to perform specific tasks efficiently and accurately.[6]

Life's processes at the molecular scale are depicted here as carried out by molecules performing technological jobs like machines—there seems to be an inventory of life forming a universal toolbox for carrying out physiological processes. For example: Specific enzymes, i.e., proteins performing specific biochemical reactions, help catalyze the copying of a DNA-strand by linking its components, similar to a tape recorder or an assembly line, whereas a rotating protein device sitting in the cell membrane produces the universal energy currency of the cell, ATP (adenosine-triphosphate; biochemists call this protein the ATP-synthase, see below), while receptor proteins detect chemical or visual stimuli. Copying machines, molecular "turbines" or "motors" as well as "switches"—these and other exemplars of protein machinery are found in every organism, or so the reader is told. Goodsell speaks of a "common birthright of molecular machines"—as this protein machinery is coded in our genomes, it represents a widespread evolutionary heritage (in fact, human beings share much machinery with microbes).[7] This also means that molecular machinery forms a common basis of explanation in very different branches of the life sciences providing causal explanations, from plant physiology to microbiology to biomedicine, from fundamental to applied.

How did life scientists explain what was going on in our bodies before the molecular-mechanical vision became as dominant as it is today? An answer to this question is difficult, as it would have to include very different types of explanations on different levels—for the case of a heartburn, doctors may have stated that cells of the gastric mucosa secreted more acid, zooming out a little from molecules to tissues. Thus, many biological phenomena simply had no molecular explanation. In other cases, such as regarding enzyme function, different models existed, for example in colloid science, which focused more on small molecules than on large, and more on chemical reactions than on mechanical processes such as movements (see chapter 3).

The question of how science was able to zoom in as close as it did on biomolecules in past decades, and how molecular mechanisms became so dominant, brings us right into recent history, and thereby into what this book will explain. Goodsell's descriptions and especially his images of "working" biomolecules resulted from insight into the spatial structure of these, and until a few years ago most of these insights were based on models from X-ray crystallography, a method to obtain data on which atom sits where in a molecule by exposing it to radiation and reconstructing its spatial structure from the spots this produces on a photographic plate. Everybody has seen such models of molecular structure—the most famous is the double helix of the hereditary substance. Since the DNA days of Francis Crick, Rosalind Franklin, and James D. Watson more than sixty years ago, a plethora of similar structures of proteins have come about, such as from hemoglobin, the stuff that makes blood red and transports oxygen through the body. Crystallographic models of DNA and protein structures have formed one mainstay of how life's makeup and processes were explained by postwar molecular biology, and, not least, images of these models have become icons of scientific and biomedical progress—countless logos and sculptures of the double helix bear testimony to this.[8]

The inventory of enzymes, structural proteins, and functional molecular complexes whose structures have been "resolved" (the scientists' term for obtaining a spatial model of an unknown molecule) is ever increasing, and novel methods to get such structures (such as by electron microscopy [EM], discussed in chapter 2), of building and displaying these models, e.g., on computer screens, and of working with the models have been conceived since the 1960s.[9] Repositories filled with the data on a plethora of DNA and protein structures from different organisms and imaged under different conditions, nowadays online databases, allowed Goodsell to depict manifold scenarios of how life works on the molecular scale in the first 1993 edition of his book.[10]

Let us look more closely at one example for Goodsell's book to understand what message these images convey, and what their implications are: Plate 1 is a rendering of a portion of a bacterial cell as an assembly of molecules, from the yellow-reddish threads of DNA and the transcription machinery (DNA polymerases) toward the cell's center on the lower left, to the cytoplasm, in blue.

This latter is depicted on the molecular scale, not as a drop of watery solution as one may imagine, but as a space that is quite crowded with, for example, the protein-making machinery of the ribosome (i.e., the "tape recorder," in purple). The cell and its outer membranes are shown at the upper right (here, rendered in light yellow and green) as a curved film forming a boundary between the cytoplasm and the exterior; finally, there is a hair-like sugary coating protecting the cell from the outside. Whereas the yellow parts of the membrane make up the lipid film that forms the actual boundary between the interior and exterior (not unlike the delicate film of a soap bubble), the greenish blobs within represent the kinds of "pump" and "channel" proteins—and this book will describe the history of research on this type of machinery. The plethora of analogies to devices from our macroscopic world notwithstanding, Goodsell argues that the molecular machinery performs "their jobs in a strange, unfamiliar world"—for example, they are driven by random molecular vibrations or Brownian motion, that is, they bump around until they find the right place.[11] Still, the strong resemblances between the mechanical, machine-like action of molecules and macroscopic devices are evident.

This book's leading question will be to find out precisely how, in the last quarter of the twentieth century, the life sciences came to consider cells and their substructures as such "molecular landscapes," i.e., as ordered arrangements of molecular machinery. I will address this problem not primarily by following problems related to imaging and modeling techniques (although this is done in chapter 2 for a novel electron microscopic approach). Imaging and modeling in X-ray crystallography has been fairly well studied by historians of postwar molecular biology, such as in Soraya de Chadarevian's monograph about "structural biology" (thus the name of the field in between physics, chemistry, and biology carrying out X-ray crystallography and related methods) at Cambridge's LMB.[12] Instead, I will focus on how these mechanical molecules have *materialized* in the laboratory—by isolating proteins from organisms biochemically, by modifying them with the help of chemical methods or putting them together in a "plug-and-play" arrangement that displays their function in the test tube, or by attempts to actually turn them into devices. My assumption is, that such practices addressing proteins as the active "springs of life" (to adapt a term by Evelyn Fox Keller) have changed the latter's materiality since 1970, leading to a situation as described at the outset, where it appears self-evident and daily practice to block a molecular pump within a stomach.[13]

Goodsell has also hinted at the personal and social implications of the molecular-mechanical vision for how scientifically informed contemporaries conceive of and act upon life under the heading "You and Your Molecules":

> Your molecular machines are far too small to see. You might think that it would be impossible to affect them yourself, to speed them up or stop them, since they are so tiny and inaccessible. However, we modify the action of our own molecular machines every day. If you take a vitamin each morning, you are tuning up your molecular machines, making sure they are in top form. If your doctor gives you penicillin, you're actively attacking the molecular machines of the bacteria in an infection. [. . .] If you take an aspirin, you are blunting the function of molecular machines in your nerves and brain. With vitamins,

poisons, and drugs we deliberately modify the action of specific machines, and by careful use, we can improve their action and thus our own quality of life.[14]

This statement adumbrates a materialistic perspective on life, which takes health a matter considered not so much at the level of a society, the psyche, or even a whole organism, but of its components, and thereby of specific portions of active matter that can be modified, even "improved." So, in a broader perspective, contemporary discourse on molecular machinery in science and medicine forms part of what Nikolas Rose has termed "molecular biopolitics," that is, ways in which molecular elements of life become modified, mobilized, and transformed in novel ways in order to alter bodily states, with the ultimate aim to change or to optimize the self. With respect to the neurosciences, for example, Rose and Joelle Abi-Rached speak of the contemporary "neuromolecular gaze," as a common language, ethos, and approach brought about by the practices and techniques of this field—the examples of Prozac but also of heartburn pills seem a prime example for what this means within and beyond the world of "neuro."[15] One of the many pertinent historical, sociological, and philosophical questions that will not be addressed in this book is *why* (post)industrial societies at the turn of the twenty-first century have tended to predominantly conceive of their health or other bodily processes and states in such molecular-mechanical terms, as opposed to explaining them by reference to higher levels of biological organizations such as organs or organisms and their constitution, or even by social factors such as nutrition or stress. The issue of why "somatic" individuals, as Rose has called this way of conceiving of bodies, and health in the age of biomedicine, have become part and parcel of our expert and everyday cultures would need to go into analyses of medicine, healthcare systems, the pharmaceutical industry, politics, selfhood, etc.[16] Here, I will restrict myself to understanding how this novel molecular biopolitics has been put into place, or "realized" in a material sense, by transforming the materiality of life in the hands and minds of scientists in the laboratory. This may be important to stress, since this book is not primarily a study of contemporary biomedicine, but of fields of the life sciences and notably also the chemical sciences (membrane and protein research, organic and colloidal and biological chemistry) which have been influential in shaping the current molecular-mechanical vision in biomedicine and elsewhere, and which allow an exemplary historical investigation of how we got to the point where we are. As the history of materials, instruments, and scientific practice described in this book will show, science, technology, and medicine of societies around the globe have become able to act with and on our bodies, as well as on living beings in and around us, in this particular molecular-mechanical fashion. Insofar, this study should be considered as a stepping stone on a terrain that has found hardly any historical attention, and that invites other studies bringing in, e.g., political and economic developments related to health, medicine, and technology in recent decades. Let me explain another central aspect of molecular machinery. So far, the image we have encountered of the molecular microcosm has been essentially static. However, when thinking of any physiological process—be it the blocking action of a proton pump inhibitor, a protein motor, or the transfer of a signal through the cell membrane—we must imagine dynamic molecules. To this end, scientific illustrations include a series of snapshots, creating something like a cartoon strip, or videos displaying simulations of molecular dynamics over fractions of a second. Such visualizations are frequently found in scientific periodicals or online tutorials nowadays. In 2000, *Nature* featured a "News and Views" piece under the title "Bacteriorhodopsin—the Movie," illustrating the detailed functioning of a model molecular machine, the membrane-bound proton pump bacteriorhodopsin (BR; fig. 1).[17]

Here, this pump's transport of a proton across the membrane was displayed as a series of mechanical steps. The "tilting" or "bending" of the protein's parts, modeled simply as seven rods stretching across the membrane, mechanically pushed its freight, the proton, across the boundary of the cell. The umbrella term in science for such shape changes or molecular movements of proteins is that of "conformational changes." Sequences of coordinated conformational changes, e.g., movements of the protein's rod-like subunits in the present case, explain its biological function. Conformational changes make the analogy between proteins and mechanical devices quite strong and seemingly self-evident—just like in a macroscopic mechanical (Cartesian) machine, internal movements of clearly discernible parts of a molecule bring about its function. In other words, the abovementioned molecule is considered a pump since there are mobile elements in its organization that push something over a distance, in this case moved by the energy of light. Note also what is discussed in more detail below, namely, that this molecule is thought to be endowed with a specific activity related to life, whereas at the same time it could also be regarded as a complex, structured, and reactive form of matter.

In addition to images, many professional articles from the contemporary life sciences propose quite extensive—for an outsider tedious and

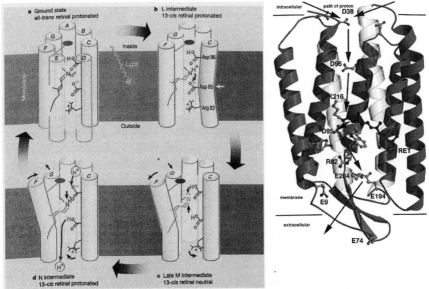

FIGURE 1 Scientific representations of proteins as molecular machinery, c. 2000. Left: Cartoon of the mechanism of the protein bacteriorhodopsin (BR). Illumination of the protein, schematized as seven rods (known as alpha-helices), leads to a series of "conformational changes" (twisting and bending of the rod substructures) that catalyze the transport ("pumping") of a proton across the membrane (the central dark region). At the end of the catalytic cycle, the protein returns to its initial state. Right: Computer generated model of BR's structure, showing the same process at higher resolution. Such molecular models, generated on the basis of data indicating the position of atoms in space, are the central result of X-ray crystallographic analyses, as pioneered in the 1950s on DNA (see chapter 1 and 2 for more historical detail). The protein's rod-like alpha-helical elements are depicted schematically as spirals, or as flat arrows (beta-sheets). Functionally important amino acids of BR are highlighted as ball-and-stick models. Arrows indicate the path of a proton from the intracellular to the extracellular space (top to bottom). The cell membrane is indicated by black lines. Left: from Kühlbrandt, Werner. 2000. "Bacteriorhodopsin—the Movie." *Nature* 406, p. 569; Right: courtesy of D. Oesterhelt, Martinsried. Reproduced with permission.

incomprehensible—descriptions of what causes what within a molecule. These mechanical narratives, with actors being molecules or their parts, explain the function of a protein at the level of its substructures in a mechanistic framework, down to single atoms. Here is an example to skim over from a 1998 publication detailing the steps of BR's "transport mechanism":

> After photoisomerization of the retinal from all-*trans* to 13-*cis*, 15-*anti*, the Schiff base proton is transferred to Asp-85 located on the extracellular side, and a proton is then released to the bulk from a site near the surface. [. . .] Reprotonation of the Schiff base is from Asp-96 located on the cytoplasmic side, aided by tilts of the cytoplasmic ends of helices F and G that were

thought to result in increased hydration of this region. Reprotonation of the Schiff base through a proposed chain of water
molecules is followed by reprotonation of Asp-96 from the cy-
toplasmic surface and reisomerization of the retinal to all-*trans*.
Finally, transfer of a proton from Asp-85 to the vacant proton
release site completes the cycle.[18]

Forget all the incomprehensible details of the process, graphically displayed in figure 1, whereby this protein, sitting in the cell's membranous
boundary, transports its freight by way of rearrangements of protein
structure out of the cell, and returns, after a series of specific conformational changes (transfers, tilts, isomerizations, etc.) to its initial state,
thus completing a functional cycle. The point to keep in mind is that
in opposition to mathematical expressions of physics or theoretical biology, or to chemical formula in a reaction equation, the explanation
given here takes the form of a highly complex narration. For each protein, the respective experts could tell the function in similar ways, identifying crucial steps of functional cycles, or important elements, such as
a channel's "gate" or "hinge," and of course it is these explanations, in
conjunction with images, that allow scientists to conceive of ways to interfere with this process—metaphorically speaking by throwing a span-
ner in the molecular works. For their complexity of molecular cast and
action, and the descriptions of the process in "scenes" of a "play," recent
systems biologists have called this explanatory style fittingly a "Shakespearean biology."[19] Whereas this label may have been intended to be
slightly deprecating from their point of view, as systems biology aims at
a "Newtonian" mode of explanation, in which processes of life are to be
spelled out in mathematical language, Shakespearean molecular narratives excite many biochemists, biophysicists, and drug designers, even
if they may sound to outsiders like descriptions of Rube Goldberg machines. Plentiful, highly idiosyncratic, detailed and branching (this was
the origin of the analogy to Shakespearean plays), similar narratives of
molecular structure and action as well as interactions are spelled out,
amended, revised, and retold in the contemporary life sciences day by
day. However, within the sheer diversity of molecular narratives there
are also recurring motives and plot structures of molecular action to be
discerned. It would be promising to scrutinize the distinction between a
Shakespearean and a Newtonian biology for an epistemology of explanations in the life sciences (notably, one would have to take into account
chemistry here, and its descriptions of reaction mechanisms by text and
formulae). However, in this book, I am not so much interested in differ-

ent biological explanations, but in exploring how science came to conceive of such narrative descriptions of mechanisms in the past four decades by materializing molecular machinery in the lab. Which fields of the life and the chemical sciences contributed to this research, which instruments and approaches have been used to materialize proteins as mechanical devices, which conceptual changes have occurred? How have molecular mechanisms and proteins as machinery become so naturalized and self-evident that all kinds of biological processes are spelled out in this framework today, and that we can buy a proton pump inhibitor pill at a pharmacist's without wondering too much whether we *really* are composed of tiny machines?

Descartes among the X-ray machines? Mechanisms, molecular machines, and the epistemology of science

The molecular-mechanical vision as described so far has a striking look-alike in the history of natural philosophy. (In)famously, early modern natural philosophers have explained life by the motion of corpuscles circulating through the body, being moved for example by sensual perception, or flowing through nerves and blood similar to gases or liquids in hydraulic machinery. In fact, such materialist theories of life were linked to the idea that organisms represented merely mechanical automata, as modeled by Jacques de Vaucanson's famous eighteenth-century mechanical duck. Whether different versions of mechanist theories included a nonphysical spirit of human beings, René Descartes' *res cogitans*, or whether they reduced also mental states to physical events, such as in Julien Offray de la Mettrie's "Man a machine" (*L'Homme Machine*, 1748), the analogy to Goodsell's picture of life is obvious—there seemed to be not much more to life than matter, corpuscles, and mechanics.[20] So, does the dominance of the contemporary molecular mechanical vision mean that early modern mechanical theories have been vindicated? Put this way, the question is obviously misleading, but a comparison of aspects central to these accounts may help to sharpen our picture of what we are seeing today.

First of all, the early modern natural philosophers considered organisms or their organs as machine-like, whereas in our cases, the question pertains to biomolecules. The discussion whether organisms are, or can be sensibly described as, machines has been ongoing since the nineteenth century also in biology, and often enough the early modern mechanist theories have served as a background, if not a caricature, against which organicist or vitalist theorists have sketched alternatives. Still before the rise of modern biology, but inspiring its discourse, Immanuel Kant high-lighted the fact that in contrast to mechanical machines, organisms dis-played a different relationship of whole to parts and that they had the capacity to reproduce, i.e., to sustain themselves and to bring forth off-spring developing a similar organization. French historical epistemolo-gist Georges Canguilhem has developed Kant's argument further in the twentieth century, asking for a switch of perspective in which the knowl-edge of organisms, such as ourselves, should be considered as epistemo-logically primary, and mechanical machines as derived or secondary. Even if the problem of molecules as machine-like entities may differ in important aspects from the long-standing issue of organisms and ma-chines, this argument is worth keeping in mind.[21]

Second, a family resemblance between these two approaches is their focus on mechanisms to explain processes of life. In this context, it is remarkable that in parallel with the rise of the molecular mechanist vi-sion since the 1990s, "mechanism" has also become a privileged concept in the philosophy of the (life) sciences through the works of William Bechtel, Lindley Darden, Carl Craver, Marcel Weber, and many oth-ers.[22] Darden and Craver, however, seek to sharply distinguish "mecha-nisms" from "machines." They point to the terminological distinction that machines are material entities whereas mechanisms are processes, and that the former can insofar harbor one or more mechanisms, but the two are not identical. One could add with Canguilhem that machines are man-made constructions, and that their functioning contrasts to processes encountered in nature. Canguilhem went on to state that no such thing as a "machine monster" existed, and somewhat similarly, Darden and Craver contrast the messiness of evolved mechanisms with engineered machines that have clearly distinguishable parts.[23] However, these points of distinction seem less obvious to me when looking at re-cent debates about molecular machinery in science. Molecular cell biolo-gist Bruce Alberts, for example, introduced proteins in the third edition of his richly illustrated and well-known textbook *Molecular Biology of the Cell* (1994) not as "rigid lumps of material," but as possessing "precisely engineered moving parts whose mechanical actions are cou-pled to chemical events. It is this coupling of chemistry and movement that gives proteins the extraordinary capabilities that underlie all the dy-namic processes in living cells."[24] In many of the historical sources ana-lyzed in this book, the clear distinction between evolved mechanisms of nature (such as in proteins) and designed machines appears blurry: Not only do scientists often talk in a self-explanatory way of proteins as ma-chines, but they model and analyze their materializations as specific machinery (pumps, channels, etc.), they characterize distinguishable parts (switches, levers, hinges), they generate novel knowledge by describing mechanical interactions of these parts, and not least, they aim to repurpose and use this machinery to do work at the molecular scale in biomedicine and nanotechnologies.[25]

One reason for why this different relationship of mechanisms and machines in contemporary science has not found the attention it deserves may be that generally mechanism philosophers did not attribute great significance to the submolecular level of protein parts, atoms, and chemical bonds. This is somewhat of a paradox: Although the "ion pumps" and "channels" of nerves, or the membrane proteins of the respiratory chain, a central part of energy metabolism, are important examples for Craver and Darden or Weber in their conceptual sketches of mechanisms, they take cells and proteins as the lowest level of mechanistic analysis, and do not go much in detail about what scientists had to say about the latter's substructure. However, it is precisely this level that is most important to the molecular-mechanical vision as outlined above, since the submolecular realm forms a common explanatory ground, addressed by biological methods from crystallography and electron microscopy (EM) to spectroscopies, from various fields such as physiology, cell biology, biochemistry and biophysics. I will argue that this is the level where science's practices and concepts have blurred the boundaries between mechanisms and machines and that it is this level that needs to be studied historically if one wants a more articulate answer to the questions sketched above.

And yet, should one not take the mechanical, technology-inspired terminology of molecular machinery as metaphorical on a superficial level, or merely as a smart way of advertising the future benefits of biomedicine or nanotechnologies, another recent endeavor where natural or synthetic molecular machinery is promoted, wrapping biochemistry in a discourse of novelty, design, and control?[26] Thus, a skeptic may argue that molecular machines actually *are* proteins, together with DNA the other main class of substances making up biological cells. More specifically, speaking the language of biochemistry, the "machinery" would form part of those proteins that catalyze biochemical reactions, which have been called enzymes since at least the 1930s.[27] However, such dismissal of the machine analogy on the molecular level as merely linguistic would be too easy. First, as studies of informational metaphors in molecular genetics have argued, a clear separation between metaphors and "proper" scientific terms is neither possible nor advisable, since it would mean missing out on understanding the generative potential of conceptual transfers, as has been shown for the case of between text, code, and life.[28] Also in the present case, the adoption of mechanical terminology has shaped a new discourse in which the assumption of a machine-like functioning of proteins actually has found explanatory value. This is evident in, e.g., Alberts' textbook or a classic volume on X-ray crystallography, Branden and Tooze's *Introduction to Protein Structure*, as well as a plethora of papers in protein science, as discussed below: Characterizing a "switch" within a protein, or remaking it, is informative for scientists as it helps them describe physiological processes such as nerve excitation, hormone action, or photosynthesis on the molecular level, and it helps them to modify these.[29]

Biomedicine, pharmacology, and nanotechnologies setting out to create molecular devices highlight another argument for why the molecular-mechanical vision has become much more than fancy or opportunistic linguistic packaging of more orthodox scientific terms. To revisit the initial example of heartburn in contemporary biomedicine, one may paraphrase the dictum of philosopher of science Ian Hacking that if you can spray electrons, they are real, as follows: If you can block the proton pumps in your gastric mucosa and record the effect both on the level of the protein (decreased function) as much as on that of the organism (decrease of acidity in stomach, relief of pain), these pumps must have become real in some way to the scientific community, and to those endorsing its knowledge production.[30] My adoption of what became almost a slogan emblematizing entity realism, introduced in Hacking's 1983 book *Representing and Intervening* (foundational for a combined approach of history and philosophy of science, as well as of the practical turn), does not mean that I would like to make a normative, transhistorical statement on the reality of molecular machinery. By contrast, it appears to me that the process of *concretion* and *materialization* of what grew out of a long-standing cultural stock of mechanical metaphors may help us to understand the historical development not only of science's concepts, but also of its objects.[31] For comparison, remember that an object as material and as self-evident to biologists as the cell started its development as a metaphor in the nineteenth century, when a concept known from architecture and electrochemical technology was transferred into biology.[32] And there is evidence that molecular machinery acquired a similar, uncontroversial matter-of-fact status as cells have around 2000: A report in *Nature* on visualized movements of the ATP-synthase rotor protein stated that "some enzyme complexes function literally as machines, and come equipped with springs, levers and even rotary joints," whereas a historical monograph on protein science introduced the sodium/potassium pumps of the neuronal membrane (the proteins that create nerve electricity) as "robots par excellence."[33] Even a scholar of science as critical to blunt reality claims as Evelyn Fox Keller stated in a recent argument on the history of active matter that "molecular motors deployed in intra- and extra-cellular motion are interesting in part because they are so dramatic: Watch these processes unfold and you know you are watching life."[34]

Certainly, there have also been skeptical voices: Cell biologists have argued against the machine analogy, calling for some sort of "molecular 'vitalism'" on the basis of differences between macroscopic and microscopic mechanics (molecules bumping around statistically, as heard above)—however, even these authors used the term "molecular motors" affirmatively throughout their article.[35] In an intriguing book on contemporary protein crystallography, anthropologist Natasha Myers develops another angle of critique: She focuses on practices involved in the creation of molecular models and explanations, and pays specific attention to how the bodily actions of the researchers can be positioned vis-à-vis the mechanistic understanding of life that they promote.[36] She puts the all-embracing explanatory claims that the molecular-mechanical vision of life entails into perspective, particularly with regard to researchers' lab discourse, which abounds with tropes regarding teleology and liveliness (such as when proteins are said to "breathe"). To address such critique comprehensively, however, one would have to ask who advocated what type of machine concept in which context, when, and what came out of it—these are all questions that will be dealt with in detail throughout this book.

To sum up, I argue that important distinctions between early modern mechanistic theories exist not only on the level of theory and concepts, but especially regarding materialization and practice, and possibly these characteristics have contributed to making molecular machinery self-evident among many scientists—and therefore worth questioning. Based on a historical understanding of Hacking's theme, I will analyze the emergence not only of the molecular-mechanical way of thinking, but also of the corresponding ways of acting on life in the lab. I will expose why and how this vision of life has become as powerful as it is in the present-day life sciences by analyzing not so much the problem of metaphors or visualizations, but primarily the materialization of molecular machinery. That is, I will describe how a molecular pump has become real to scientists as a material substance that could be isolated from cells, that could be modeled, modified, and remade—in iconic experiments, its actions were observed with the bare eyes.

If historians and philosophers should acknowledge the omnipresence and impact of the molecular-mechanical vision of proteins, this should also not be taken as an argument vindicating molecular mechanisms as a fundamental or privileged means of explanation for the processes of life. My take in this book will be descriptive, and thus steer between the normative philosophical positions of either naturalizing molecular mechanisms or criticizing, even rejecting it wholesale, e.g., for being unable to adequately explain higher level biological phenomena.[37] I will tread this somewhat agnostic middle path, because I am interested less in the philosophical problem of explanatory adequacy and more in studying agency through historical sources. A "normative moratorium" is, I believe, beneficial to understand why and how we have reached our current position in the life sciences, considering the present neither as a corrupted state nor as the end of history (contemporary scientific insights will be discussed in the Conclusion).

Life and matter—another history of the molecular life sciences after 1970

Turning from epistemology to historiography, this book problematizes two central issues of the recent life sciences. First, the history of membranes and molecular machines can be read as a novel answer to the question of "what happened after molecular biology," that is, after the postwar molecular "revolution" of biology connected to names such as Watson and Crick. Second, and in connection to that, this story highlights the relevance of chemical objects, practices, and concepts to the development of the life sciences, revisiting received narratives and getting into view an often neglected, but pivotal dimension of these latter, that of matter and its activities.

The historiography of molecular biology after 1945 is still dominated by the narrative of the rise of molecular genetics—from the double helix to the introduction of cybernetic, informational discourse into genetics to the cracking of the code in the late 1960s, as described beautifully in Lily Kay's *Who Wrote the Book of Life?* These developments led to the first experiments on recombinant DNA, or genetic engineering, around 1970, that allowed a "rewriting" of the code, thereby sparking molecular biotechnology and consequently our genomic present. A second, smaller historiographical strand has focused on biophysical research such as in structural biology, that is, EM or X-ray crystallography of proteins. Whereas the leading background metaphor employed to understand life on the molecular level has been that of reading and rewriting the book of life in genetics, it has become that of machines and a mechanical functioning of organisms in structural biology. Thus, Soraya de Chadarevian has analyzed the emergence of molecular mechanisms between structural biologists and biochemists studying enzyme activities at Cambridge's LMB in the 1960s.[38]

This study will take up these two historiographical strands, but embed them into a picture of a molecularization of life in the twentieth century that transcends both: Membrane history, bringing in physiology, cell biology, bioenergetics and biophysics, but most importantly organic, colloidal and biochemistry, will show how this molecularization has been more than genetics plus structural biology before 1970, and how these different fields of research have been recombined in the life sciences as we know them today, which integrate concepts, techniques, and resources from all of them (chapter 1 and 2). A genealogy of practices, proposed as a historiographical approach in chapter 3, seems especially pertinent to understand the development of a seemingly very recent field, synthetic biology. Moreover, this story will complexify the picture of what happened after 1970, by revealing, to simplify quite a bit, that biotech was not only what it often appears to be, namely biomedicine and business, that the recent molecular life sciences are not only development and epigenetics or genomics and computers, and that the natural history approach of collecting genomic or structural data coexisted with a further flourishing of the experimental approach of earlier biochemistry and molecular biology. The story of membranes and molecular machines will make it clear that this approach, closely knit with the materialist and reductionist thrust of postwar molecular biology, has remained alive and well at least in some quarters of these sciences, although in a transformed way.[39]

I will address these issues by flanking my case study of the emergence of a model molecular machine by broader analyses of the membrane field before 1970 (chapter 1), by its ramifications with a synthetic biology avant la lettre (chapter 3), and by bio- as well as nanotechnological projects after 1980 (chapter 4). My case study on the emergence of a novel research object connects to Angela N. H. Creager's *Life of a Virus* (2002); she has followed tobacco mosaic virus as a model research object within different sites and cultures of biochemistry and molecular biology from the 1920s to the 1960s. Another important point of departure to conceive of the coming into being of a new research object has been Hans-Jörg Rheinberger's *Toward a History of Epistemic Things: Synthesizing Proteins in the Test Tube* (1997). This book's argument about the emergence of novel "epistemic things" in the course of laboratory bench work has been particularly germane to reconstruct at the microlevel what happened in membrane research around 1970. Somewhat similar to the "conjunctures" of material preparations with functional tests described by Rheinberger, the molecular machine in the center of this book also took shape by relating materials in test tubes to molecules with a defined structure, and ultimately to biological function (chapter 2). Creager's and Rheinberger's work on viruses and protein synthesis, respectively, are important for another reason: By taking into account chemical techniques (e.g., purification, preparation) and concepts (e.g., analyses and synthesis), they reveal that often enough, biochemistry and molecular biology have represented rather two sides of one coin than separate disciplines or fields. Take the case of how cells make proteins: One could frame this either in terms of molecular biology as a process of information flow from DNA to RNA to protein (Crick's "central dogma") or as a biochemical synthesis of a molecule within a cell. It is important to stress this ambiguity since it shows that the underrepresented chemical dimension of molecular biology is already there— inherent in the sources and to be picked up from existing historiography by reading them differently.[40]

However, in comparison with Rheinberger, there is also an important difference in how this story conceives of scientific objects and their development. Here, it is less about materials that exist as mere "traces," as minute portions of substances analyzed by instruments, but as almost brute matter on hands: Stuff as evident as "a slap in the face."[41] Thus, somewhat similar to the materials in the earlier days of chemistry the reactions of which sometimes hit back at those investigating them (in the worst case by blowing them up), the materials and processes analyzed in this book subvert the boundary that Lorraine Daston has drawn in between mostly elusive and hard-won scientific versus quotidian, self-evident objects. It is the peculiar materiality of objects of chemical research, a discipline characterized by Ursula Klein and Carsten Reinhardt as a "materially intervening, productive enterprise changing nature, technology and society" that is at the heart of this story, and the relevance of which for the recent life sciences I intend to flesh out: The making and the impact of matter's macroscopic aspect, of that which can be stared at or touched, thereby providing access to an apparently ineluctable level of reality, informs my take on the problems of materiality in scientific practice.[42] As I will argue, such *tangible* properties of matter and thereby objects of research have played a far larger role than one may expect in the age of the highly instrumented life and chemical sciences. To conceive of what appears as an ordinary but overlooked trait of matter as literal *stuff* that "smite[s] the senses," I will introduce the German term *Stoff*, designating both material substances and textiles or fabric, that is, something produced and structured that lends itself to usages (chapter 2).[43]

More specifically, the matter that stands in the center of this book's narrative can be characterized as a structured biological molecule endowed with specific activities (i.e., a machine). Thus, it is a prime example of what has been recently discussed by scholars of science and technology as "active matter." As a concept originally from physics, active matter obviously includes many entities and phenomena displaying coordinated behavior, enlivened or not, on the micro- as well as the macrolevel (flocks of birds have been mentioned as much as nanotechnological materials changing their properties). Active matter has challenged the stereotype of matter as inert, inherent also in the early modern mechanist theories, and, one should add, as homogeneous.[44] From what has been said about membranes, molecular machinery, and its materializations, it should be clear that the case discussed in this book provides one step toward a history of active matter research avant la lettre. Indeed, many episodes from the history of biochemistry (enzymes) or surface chemistry (membranes, films) of the interwar period (chapter 1), bioenergetics and biophysics (chapter 2), material cell models (chapter 3) and finally early nanotechnologies tackling material activities (chapter 4), form a strand of inquiry into peculiar, lifelike activities of matter, which Evelyn Fox Keller has followed back to nineteenth-century research on protoplasm. Whereas debates on this topic, as much as on the activities of organisms in general, sometimes suggest vitalistic, or even hylozoistic explanations, this story will exemplify what historian of science Bernadette Bensaude-Vincent has demanded: A closer look at chemistry and its concepts related to matter, such as molecular structure and reactivity, will clarify how matter has been conceived of as active, and thus situate this topic within the history of the sciences.[45] In contemporary debates on active matter in between architecture, materials science, biotech, and computing, the idea of "programming matter," i.e., of creating self-organizing, sensing, or reacting materials, looms large.[46] The story described in this book provides a concrete historical case in point of an early endeavor to put matter to work, the expectations and difficulties experienced by the actors, as well as the oscillation of this project between the mundane (chemical synthesis, chapter 3) and the quixotic (the biochip, chapter 4).

As this book highlights the relevance of chemistry—as the prime science of matter and material activities—to the recent life sciences, it can also be read as a response to Angela Creager's recent "chemical reaction" to the historiography of this field.[47] Taking up her pun, my hope is that this history will be "reactionary" by reemphasizing the influence to the late twentieth-century life sciences of fields such as biochemistry, organic or colloidal chemistry, which have been studied by a previous generation of historians, including, e.g., Frederic Holmes or Robert Kohler, but the relevance of which has been overshadowed by focusing on molecular genetics and genomics in the past decades. This is not to revitalize the old debate whether molecular biology or biochemistry was more important—as Lily Kay has argued long time ago, the rise of the molecular vision of life since the 1930s was linked to no small degree also to the "protein paradigm" of the gene, and thus chemical thinking and working, before DNA moved center stage in the 1950s.[48] Instead, I intend to reactivate the historiographical as well as the epistemological potential buried in this older historiography by following its themes into a later period and thus by using it to ask novel questions in a different framework. In this sense, I hope to be not only reactionary, but *reactive* in the sense of bringing together existing studies on these topics in this book, which might hopefully create a critical mass, and this will be the last of my stolen chemical puns, to create a chain reaction for further studies on similar subjects. To oversimplify dramatically: This story can read as an argument to challenge the Linné-Darwin-Mendel stronghold in the history of the life sciences, and to resurrect Lavoisier, who has started to frame essential functions of life, such as respiration, in terms of material processes and chemical reactions. More seriously, it should be read as an argument to forget about founding fathers altogether, and to confront the diversity and heterogeneity of recent science, asking new questions pertinent to our day.

Constitutive and exemplary: Bacteriorhodopsin, membranes, and the rise of molecular machinery

The history of life's molecular machinery could be written by looking at various domains of cell biology, biochemistry, biophysics, or molecular biology. Readers familiar with postwar molecular genetics may immediately think of the ribosome, a complex of RNA and proteins catalyzing cellular protein synthesis, which has been analogized to an assembly line, a tape recorder, and other informational devices.[49] Another example mirroring the influence of cybernetic discourse on molecular biology would be DNA polymerases, the enzymes catalyzing the fast and accurate copying of the double helix's nucleotide strands. The American biochemist Arthur Kornberg, whose name is synonymous with decade-long work on polymerases, lovingly called them "astonishing machines of replication."[50]

In this book, I will explore the history of molecular machinery by studying the history of research on cell membranes. There are reasons for writing an entangled history of membranes and molecular machinery: First, some prime examples of these machines, the ion pumps or channels, the hormone or light receptors, or the photosynthetic reaction centers mentioned above have been found in membranes. Their activity, the enigmatic physico-chemical processes of specific membranes, have puzzled scientists (as documented in the expression of the "riddle of surface action," see chapter 1) and attracted scores of plant or neurophysiologists, biochemists, and many more since the late nineteenth century to the study of what they called, e.g., "excitable tissue." While such membrane research had a first heyday in the interwar period, it remained scattered and largely stagnant after 1945—a molecularization comparable to that of genetics set in only after 1970, and it quickly led to the characterization of said pumps, channels, and transducers, and the spelling out of their molecular mechanisms. Thus, writing what I consider an exemplary aspect of the history of biological membrane research, which is a much broader field encompassing cell biology, immunology, physical chemistry, etc., allows me to analyze the rise of an important part of the molecular life sciences as we know them today, and a constitutive case for the rise of molecular machinery.

Even if membranes remain as yet largely a desideratum of scholarship, a number of studies exist—among them Max Stadler's dissertation on cells and membranes in between colloids, physical chemistry, and neurobiology during the interwar period, Daniel Liu's work on microscopic techniques leading to the lipid bilayer model of the membrane, Kärin Nickelsen's monograph on photosynthesis (another membrane-bound process central to biology) and several, mostly philosophically inclined studies on bioenergetics, i.e., the postwar research field setting out to find molecular explanations for how cells generated energy at their membranes.[51] My hope is that this book serves as a platform bringing these and other studies together with a large number of primary sources in order to shape a first, preliminary picture of another thorough molecularization of life after and beyond molecular biology (conceived of, following Robert Olby, as genetics plus structural biology).[52] Arguably, other times and other cases may modify this picture considerably, but it appears to me that this book comprises many leads that could be followed in one or the other direction.

A bibliometric analysis of key terms in scientific publications provides tentative quantitative evidence backing up my assumption of a joint rise of membranes and molecular machinery after 1970, as well as of the constitutive and exemplary character of my case study on the purple-red membrane pump BR. Researchers in the molecular life sciences and adjacent fields increasingly used the terms "pump" or "channel" in conjunction with "membrane" in their publications throughout the period analyzed by this book (c. 1965–1990), with an even steeper increase in the 1990s, when the molecular life sciences in their present state began to unfold.[53] Concurrently, the frequency of the terms "molecular mechanism," in combination with "protein" or "membrane," began to rise from the mid-1970s through the 1980s, and continued to increase at a pronounced rate after 1990.[54] These developments correlate to the drastic overall increase in publication number on BR after 1975 (peaking around 2000); moreover, also within this corpus of papers, the frequency of the term "mechanism" increases after its first usage in 1973.[55]

The generic term "molecular machine" used in combination with various proteins first appears in scientific journals the 1990s, when more proteins became characterized on the molecular level, and as increasing numbers of protein structures accumulated (see chapter 4 for the appearance of this term in popularizing journals). Two landmarks of this development may have been the Nobel prize for the first membrane protein structure, a photosynthetic reaction center, in 1988, and for the molecular rotor of the ATP-synthase in 1998—the latter called by the awardee, the American bioenergeticist Paul Boyer, a "splendid molecular machine."[56] The frequency of the term "molecular machine" further increased in the new millennium, presumably in conjunction with nanotechnology (chapter 4).[57]

A note on people and places, times and sources

This history centered on materials and objects can also be read somewhat against the grain, as a rapprochement to the people, the places and the institutions, or the times or places of a recent and only partly studied phase in the history of the life sciences. There are certainly limits to such an understanding, as the archival records I have studied, as well as the interviews I have conducted with a number of important actors, are confined to the geography of research of a case study. Centers of the BR project existed, among many other places, at the University of California at San Francisco (since 1969), the LMB at Cambridge, UK (since 1973), and the University of Munich or the MPI of Biochemistry in Martinsried, Germany (since 1970). However, the large number of published primary sources analyzed in this story has enabled me to considerably broaden and back up a case story that is already intrinsically transnational, and which includes a number of countries novel to the map of the molecular life sciences, such as the Federal Republic of Germany (West Germany hereafter) or the Soviet Union.

Chapters 1–3 are largely based on a wide range of sources of academic science from notebooks to patents, or from scientific articles and monographs to correspondence between researchers, grant applications, pedagogical material, or graphic representations and images. Chapter 4 extends the focus by bringing in popularizing magazines, books, newspapers as well as "science journals beyond the scientific journal" presenting the topic of molecular technologies to specific target audiences or a general public. These sources bear testimony to an important shift of the discourse on life and technologies in the 1980s that was possibly related to scientists' self-understanding and expectations. A number of research interviews with relevant actors have been carried out to complement textual sources. These retrospective narrative constructions of both the material and the social dimensions of the actors' research have supplied information that is rarely put on record in the sciences apart from recollections, e.g. regarding mutual perceptions, personal attitudes on certain issues, or memories of important moments. These interviews, in combination with a number of other written or oral exchange with actors, have served as a heuristic guiding the investigation of written sources (e.g., laboratory notebooks, popularizing magazine articles) rather than as oral histories in themselves.[58]

In the Conclusion, I will provide some leads as to how this book could also be read as providing insight into the specificity of the two central sites, San Francisco and Munich—hubs of high-tech in their respective countries—as well as to the encountered persona of the life scientist in the 1970s and 1980s, in between research and biotechnology.

Outline of the book

Chapter 1 will provide basics on biological membranes and the history of research on their structure and dynamics from the interwar period until 1970. I will familiarize the reader with basic concepts and experimental approaches, sketch the field's connections to (neuro)physiology, bioenergetics, enzymology, cell and structural biology, and outline why membranes should be of interest to historians and philosophers of the life sciences. The bottom line of this chapter is that membranes have remained since their first heyday in the interwar period at one remove from providing molecular mechanisms, couched in between hope of finding a new general molecular principle of life comparable to DNA, that might explain, e.g., signal transmission or energy generation of cells, and the experimental difficulties of getting ahold of membranes and their mysterious dynamics. The history of Hodgkin and Huxley's action potential in nerve illustrates this as much as that of models of membrane structure.

In chapter 2, I will zoom in on the emergence of what was to become a model for a molecular "pump," the photoactive protein BR, and the emergence of this brand-new topic at what I call the "membrane moment" of the early 1970s, when lots of the problems described above were resolved at once. BR epitomizes the sudden coalescence of research from various directions around a materialization of a membrane and its active components—this was active matter as tangible as it gets. Using text and images from laboratory notebooks and publications, supplemented by interviews, I analyze the trajectory of the formation of a research object that started with the observation of a strikingly colored, active material substance in the test tube, that lent itself to all sorts of experimental approaches, thereby drawing in diverse researchers to a new topic and spreading globally. This chapter detailing how a molecule became conceived of as a pump can be read in many ways, for instance as a history of active matter in the molecular life sciences or as an important episode toward optogenetics, a current method using this molecular machinery to influence the neuronal activity and behavior of experimental animals in stunning ways (see also chapter 4). Moreover, chapter 2 foreshadows a development that became influential in the present, by telling the story of how a novel method to image molecules, crystallographic EM, unfolded through work on BR. In 2017, Richard Henderson from Cambridge's LMB shared a Nobel prize for contributions to this method. Through a close-up on Henderson's work, as well as on that of bacteriorhodopsin's other central protagonist, Dieter Oesterhelt from the MPI of Biochemistry at Martinsried, this chapter provides insight into the work style and the topics of a new generation of researchers in the 1970s, which formed a novel network at lesser and well-known sites of the molecular life sciences.

The tandem of analysis and synthesis, as a central trope of how chemists have inquired into matter and molecules by taking them apart and putting them back together since the nineteenth century, has inspired the overall structure of this book: Whereas its first two chapters focus on how membranes have been materially taken apart and a molecular pump has been isolated, the following two chapters take into view ensuing attempts to remake membranes and to assemble molecular machines—following a material version of Richard Feynman's dictum that one does not understand what one cannot create, and thereby illustrating the path of this research into bio- and nanotechnologies. Chapter 3 details the practices of (re)making biological molecules, as well as material models of cells (membrane-ensheathed droplets called "liposomes"), and putting such components of life back together in what I call a "plug-and-play" mode of research. From the second half of the 1970s onwards, methods of recombinant DNA, but also earlier strategies of organic chemical synthesis, have changed the materiality of life's inventory. My analysis focuses on the work of Har Gobind Khorana, a chemically minded molecular biologist inspired by syntheses, who set out to gradually remake life's components—not least the first functional synthetic gene, before he transferred this approach to membranes and proteins. His research will also illustrate how mechanisms have been spelled out by taking apart, modifying, and remaking molecular machinery. On a more general level, my focus on practices of making things biological will allow me to conceive of a genealogy of synthetic biology. The history of this recent endeavor looks quite different if we take into view seemingly mundane, chemical approaches such as the making of liposomes, material models of cells pioneered by Cambridge biophysicist Alec D. Bangham and Cornell bioenergeticist Efraim Racker, or if we reflect on the impact of machines to produce proteins and DNA (such as the solid-phase peptide synthesis automat devised by Rockefeller chemist R. Bruce Merrifield).

In the climate of burgeoning 1980s biotechnologies and microelectronics, the idea of molecules as machines moved beyond the confines of traditional labs and academia when a motley group of physicists, chemists, genetic engineers, and tech enthusiasts attempted to devise and build "biodevices" or "biochips." Chapter 4 tells the curious story of such attempts to turn a molecule into a technology that was to lead to improved, because smaller, more efficient and lifelike computing. This adds an unexpected dimension to the history of biotechnologies: Beyond venture capital and biomedicine, I will show that 1980s biotech also set out as a quite radical attempt to redesign existing technology through inspiration from life—this included a countercultural figure such as Lynn Margulis, as much as later nanotech poster-child Eric Drexler, and many other actors defying received categories of science's and biotech's historiography. Projects aiming to tackle life's molecular machinery existed, e.g., in US labs and start-ups as well as within the more conventional

German chemical industry. This chapter also forms a bridge in between the historiography of the life sciences and that of materials science and nanotechnology: My case story of a BR-based biochip will put more flesh on the bones of nanotech history, often centered on programmatic or singular individuals, by following the attempts to materialize such a technology. This chapter is also a history of changing scientific media and perceptions of science, as it follows molecular machinery from the scientific press into novel tech-zines such as *Omni*, and, as one could say, from laboratories in California to Californian labs on both sides of the Atlantic. Finally, this chapter will juxtapose 1980s biochips with optogenetics, a recent endeavor to make molecular machinery work within organisms to modify their behavior, to address neurological diseases, or to create semi-organic prostheses.

In the concluding section, I reflect on how this book's history can inform our understanding of materials, *Stoff*, and active matter in between the chemical and the life sciences and what we can conclude on the thorny philosophical question about the "reality" of molecular machinery (following a historicist understanding of Hacking's theme). I will also position membranes within a broader history of the recent life sciences, formulating a number of questions for further research that this book has opened out—not least on the geography of the molecular life sciences, and the question of how to characterize the persona of those individuals we will observe taking life apart and repositioning it to mechanisms, molecules, and active matter.

Part One
Taking Membranes Apart, Isolating a Molecular Pump

1 What Membranes Can Tell a Historian and Philosopher of the Life Sciences

Just as chemistry could not have developed without test tubes to hold reacting substances, so organisms could not have evolved without relatively impermeable membranes to surround the cell constituents. [. . .] It can truly be said of living cells, that by their membranes ye shall know them.

E. Newton Harvey, foreword to Davson and Danielli,
The Permeability of Natural Membranes, 1952.

Trivial but true—life takes place in containers. As much as multicellular organisms are surrounded by a skin, an integument, an epidermis, or other surface layers, every cell is surrounded by at least one layer of membrane, and often several of them. Thus, in most definitions of life, the container, or the boundary, seems the least controversial point, as compared to, for instance, metabolism or a system of heredity.[59] The obviousness of a boundary required to separate cells from their environment, thereby creating a milieu, a space for metabolic reactions, and a compartment that harbors hereditary molecules, may have contributed to the neglect of what specifically biological membrane research has told us about life and what it has allowed us to do with it. However, without such membranes, cells would not be discernible, and we would be left with some ill-defined protoplasm, primordial soup, or surface in which metabolism and heredity take place. Stanislaw Lem's science fiction novel *Solaris* (1961) impressively shows how difficult a scenario of life without boundaries may be to imagine. Here, an ocean covering an entire planet seems to be "alive," in the sense that it somehow interacts with the humans that approach it in their spaceships; however, within the enlivened ocean, no boundaries of an organism-like structure are discernible, leaving the psychologists and cyberneticians in utter incomprehension regarding their counterpart.

In addition to their obviousness, another reason why membranes have been neglected by the history and philosophy of the life sciences, in spite of their indubitable relevance to science, may be the fact that research into them has been much less disciplined than, say, the study of heredity in genetics. Plant physiologists of the nineteenth century stumbled upon surfaces and interfaces of tissue when studying water transport, that is, the curious phenomena of osmosis (the tendency of water to flow across semipermeable boundaries from solutions of lower to higher concentrations of ions). Everybody has experienced osmotic effects, for instance, in the miraculous recovery of withered plants upon watering, or the swelling of a gummi bear in a glass of water. Osmosis illustrates another almost trivial necessity of living in a container: As much as a boundary is required for life, this boundary must be traversed in order for metabolism to occur. Water needs to flow into plant tissues, while cells need to take up nutrients and excrete waste products, etc.[60]

Another important research field where membranes moved into the focus of early twentieth-century biology was the physiology of "excitable tissue," such as nerve or muscle. Just as in electrochemical devices (e.g., batteries, composed of two compartments filled with different ionic solutions, separated by a semipermeable membrane), nerve membranes were the site at which the electricity of the tissue was generated, which was a centerpiece of sensory physiology since the nineteenth century. Moreover, biological membranes seemed to be examples of an "active surface," similar to the artificial membranes used in, for example, filtration. Just as the thin layers of synthetics such as collodion, they were endowed with properties perceived as extraordinary: Surfaces behaved very differently than expected from the physics and chemistry of bulk matter, such as common gases, liquids, and solids. Thus biological research successfully cross-fertilized with other fields interested in what British colloid chemist Sir Frederic Donnan evoked as a "fourth state of matter" in the interwar period, among them even electrical engineering.[61]

One major step of twentieth-century membrane biology was to distinguish the biological membrane proper from other surface layers of cells and tissues, such as the cell walls of plants, fungi, or bacteria.

Polypeptide chains
in β-form

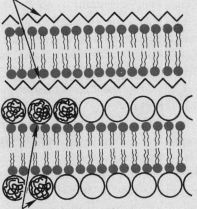

Polypeptide chains
in globular form

A bilayer model
with the polypeptide
chain in the
hydrocarbon phase

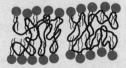

A bilayer model
with penetration of the
polypeptide chain

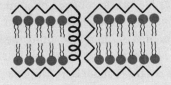

A globular model
in which polypeptide
chains coat lipid micelles

FIGURE 2 Elusive boundaries—models of biological membranes, c. 1970. This figure, taken from the first edition of Albert L. Lehninger's biochemistry textbook, displays competing models of membrane structure, illustrating the lack of consensus about membranes as a fundamental principle of life organized on the molecular level. Lipid molecules are displayed by their molecular heads and tails, proteins as saw-tooth, spiral, curvy, or circular lines (illustrating possible different conformations). Some of these competing models, possibly applying to different membranes (e.g., intracellular), had been around since the interwar period, such as Davson-Danielli type models (top; also called unit-membrane in the postwar period). For more on these iconic models, see Liu (2018). From Lehninger, Albert L. 1970. *Biochemistry: The Molecular Basis of Cell Structure and Function*. New York: Worth Publishers, p. 212.

Whereas the latter are more or less inert sheaths rendering the cell with mechanical stability (e.g., against "explosion" due to the osmotic influx of water and swelling), the membrane was conceived of in the 1930s as a delicate double lipid, or fat layer, only several nanometers thick (fig. 2).

This lipid film, too thin to be visible under optical microscopes, but intriguing for its effects, actually led to a veritable membrane-craze in the 1930s and 1940s (not only amongst neurophysiologists), before molecular biology as we now know it had formed.[62] Yet, the exact molecular architecture and the function of the thin double layers composed of lipids and proteins remained controversial even in the postwar age of the electron microscope, as the following section shows. How membranes achieved their remarkable effects—the generation of action potentials in nerve, the selective uptake of ions or nutrients in blood cells or bacteria, respiration, or photosynthesis (all of these centerpieces of twentieth-century life sciences) only became noncontroversial and properly addressed after 1970, in cases such as that described in this book.

Exploring these membrane histories does not just add a novel dimension to our picture of the life sciences. The way in which researchers have conceptualized and dealt with life in membrane research provides novel insight for a philosophy interested in the concrete, as it suggests viewpoints and questions that differ from those posed by genetics and evolutionary biology. Take heredity, for example: Cells "inherit" half of their membrane (as their metabolically produced and self-maintained boundary) when dividing. Obviously, this transfer of a material structure is a form of heredity radically different from that of DNA. Or, consider membranes' dynamic mode of existence; that is, their self-organization and the fact that they remain discernible entities in spite of continuous material exchanges with the environment, with which they form a dynamic equilibrium. Membrane research is of interest to study the interplay of the life sciences with physics and chemistry (thermodynamics, reaction catalysis, etc.), especially in regard to models for the emergence of larger structures, the formation of order, or the origin of cells. It is the material modeling of membranous objects and their dynamics—from mixing lipids and water for spontaneous membrane formation, to extractions, centrifugations, syntheses of "protocells" to the study of communication between cells and interactions with their environment—that has allowed membrane research to re-formulate and re-cast many of the central issues of the life sciences. Stories from membrane research challenge distinctions such as those between the living and the unenlivened, or the "natural" and the "synthetic."

Bringing membranes into history and philosophy will not only high-light a realm in the life sciences that is quite different from the ones that have found more scholarly attention; insight into membrane sciences will also expose the extent to which the late twentieth-century life sciences have been an endeavor of chemical thinking and working; that is, of isolation, preparation, making, unmaking, and reassembling matter.

The cell's elusive boundaries and the molecular age

Excitement and disillusion were close counterparts when it came to membranes in the 1960s. Let us first examine excitement. Under the heading "Molecular biology—the next phase," Max Delbrück hit on the subject of membranes in 1968. Evoking physicist Richard Feynman's "There is plenty of room at the bottom" address, later taken as foundational charter of nanotechnology, Delbrück outlined membranes as a form of natural technology, as the cell's "chemical factories" in enzymology, or as its "surface structures" transmitting signals in the nerve fiber, which influence cellular behavior:

> On the molecular level these [i.e., the membranes'] transducer mechanisms are not understood and will constitute the principal challenge for the next phase of molecular biology. The depth of our ignorance in this area may be compared with the depth of our ignorance with respect to the molecular basis of genetics 30 or 40 years ago.[63]

Despite Delbrück's next frontier rhetoric, and his group's work on sensory physiology of simple organisms (somewhat unsuccessfully), he was not among the important membrane researchers of the time. Yet, if we take Delbrück's role as an interdisciplinary leader and community organizer with a nose for the vanguard of science seriously, his 1968 essay indicates that membranes were emerging as a novel subject of the molecular life sciences and drew attention in the later 1960s. Potentially similar in architecture and function, membranes from different cells or organisms were another candidate for a general principle of biological function at the level of physics and chemistry. At least for Delbrück, genetics cast its long shadow over this surmised next big thing of molecular biology—he framed the topic in a cybernetic discourse on information and technology that seems familiar from both the story of DNA, the genetic code, and later neurophysiology.[64] Thus, we will again encounter Delbrück, as well as Feynman's promise of molecular technology, when it comes to 1960s "transducer physiology," and biological macromolecules as "switches" or other "devices" in the 1980s (see chapters 2 and 4).

So much for membrane enthusiasm—but what was the nitty-gritty of research at the time? In 1968, a lengthy review on "Current Models for the Structure of Biological Membranes" (note the generalizing expression, as opposed to particular membrane specimen) by Rockefeller University electron microscopist Walther Stoeckenius discussed almost 300 references, only to conclude that the most acceptable model, or at least the one with the fewest counterarguments, was the "Davson and Danielli bilayer model," which proposed two lipid films with proteins attached, and which dated back to 1930s research. Two years later, the first edition of a biochemistry textbook authored by American Albert L. Lehninger (known among biochemists as "the Lehninger" to this date) openly reflected this lack of consensus on membrane structure by depicting different models of how these were made up from their components, lipids and proteins (see fig. 2).[65] The situation was no better regarding the investigation of membrane activities in neurophysiology or bioenergetics. Arthur B. Pardee, another molecular geneticist on the lookout for new subjects, stated in 1968 that the details of membrane activities such as transport of substances into cells were "completely mysterious" and that the existing "black-box approach" of physiology did not permit one to decide the central issue of *mechanisms*.[66]

Another controversial and perhaps the best-known arena of inquiry into membranes at the time was bioenergetics, or the study of cellular energy generation, which had developed out of intermediary metabolism studies and photosynthesis in the postwar period. Membranes of mitochondria and the so-called thylakoid membranes of chloroplasts were known as the sites of production of the central energy metabolite adenosine-triphosphate (ATP), yet if and how the lipid-protein-film was involved in ATP synthesis remained a matter of controversy.[67] Whereas many biochemists thought the membrane was of less importance, and looked for enzymes and intermediate reaction products, the so-called chemiosmotic model developed by British biochemist Peter Mitchell had put the membrane center stage. Similar to the semipermeable membranes used in batteries, which serve to create ionic gradients between two compartments and thereby electricity, Mitchell thought of the membrane and the ionic gradient as a general biological stratagem to catalyze and utilize surface processes, thereby revealing himself as a late acolyte of interwar chemistry in the age of molecular biology. So, conflicting evidence and clashing styles of biochemical working and thinking had turned bioenergetics in the second half of the 1960s into an acrimonious controversy. Efraim Racker, an Austrian-American biochemist from Cornell who was deeply entangled in the debates, put it thus: "Anyone who is not thoroughly confused just does not understand the situation."[68]

From these contrasting assessments, membrane research emerges as an endeavor loaded with expectations for new general insights into life, but plagued by conceptual confusion and experimental stagnation since its first heyday in the interwar period. Membranes did not live up to their promises. To exaggerate, one could say that different membranes were researched by different communities in the 1960s: The relationship of the biochemists' membrane (a lipid fraction prepared from cells) to the physiologists' membrane (the electrical effects of which in living cells were recorded on paper slips and screens) to the electron microscopists' membrane (visualized as thin dark lines in stained specimens of cells and tissues) to that of physical chemists' (e.g., assembled synthetic thin films) remained unclear, with one participant of a 1967 conference speaking of different "tribes" inhabiting "membraneland."[69]

In the remainder of this chapter, I will present this scattered landscape of 1960s membrane research, with its cornerstones such as the bilayer model reaching back into the 1930s. I will first venture into the problem of membrane structure, then on membranes as known by remaking them, and third, on their—to echo Pardee—mysterious dynamics. The bottom line of this story is that membranes have remained at one remove from providing uncontroversial models of structure and dynamics up to the late 1960s, different aspects of membrane research remained unconnected, and most importantly—the object itself materially elusive. Thus, this chapter, composed of somewhat disconnected threads, can also be read as a story of slow-moving, meandering research, which yet prepared the ground for the coming "membrane moment" of the 1970s.

Neglected dimensions: Membrane structure

Even if much remained controversial about membrane structure and assembly in the 1960s, it was accepted that these delicate films represented dynamic aggregates of relatively small lipid or fatty acids molecules. These latter comprised hydrophobic, i.e., water-adverse, tails and hydrophilic heads, thus often being depicted in a tadpole-like schematic (see fig. 2). In the cytoplasm, or watery solutions more generally, these molecules spontaneously assemble into spherical aggregates of microscopic size, not unlike soap bubbles. Whereas the lipids' water-loving head groups would face the outside solution, the water-adverse tails would spontaneously orient away from water for thermodynamic reasons. The analogy between such processes of membrane self-organization and the formation of cells, and thereby a major step in the origin of life, is easily made, and had in fact been drawn already in the early twentieth century by the Soviet chemist Alexander I. Oparin.[70]

Lipid films as flat sheaths or the spherical liposomes and micellae everybody knows from turbid soap solutions, formed part of what German colloid chemist Wolfgang Ostwald (the son of physical chemist Wilhelm Ostwald) has famously called the "world of neglected dimensions" in the early twentieth century, i.e., a realm of aggregate structures the size of which ranged in between those of the organic molecules studied by biochemists and larger cellular structures visible in the microscope. This cosmos of colloidal substances and effects encompassed enzyme action and cell structure as much as it promised to explain the properties of paint or ketchup, all of which seemed to elude what was known from the ordinary chemistry of solutes. When crystallography and the ultracentrifuge revealed that protein enzymes were not organized as aggregates in the 1930s, but as macromolecules, colloidal chemistry as a biological research program largely collapsed, yet, membranes remained a stronghold of colloidal thinking and working. They formed if not a world, then at least an *island*, of neglected dimensions next to or within postwar molecular biology.[71]

Yet, how had researchers already concluded in the 1930s that the membrane was organized as a bilayer, if these structures were too small to be visualized directly and, as aggregates of small lipid molecules, too dynamic to be analyzed by crystallography? Interwar colloidal and biochemistry had veered into the world of neglected dimensions largely by way of indirect experimentation. The idea that cellular membranes were made of *double* layers of lipid film, for example, resulted from extracting lipids from a known number of red blood cells (easily accessible material) and comparing the surface area covered by these lipids as an unimolecular film with the calculated total surface of cells.[72] A factor of two emerged, which could be interpreted as evidence for a membrane bilayer. That is, by way of chemical extraction, the invisible membranes had become a tangible material substance, which was used to re-form a secondary, macroscopic film in an instrument. The properties of this latter allowed inferences about the molecular membrane organization in vivo, which were only directly observed years after the war. At the same time, the concept of molecular orientation or "anisotropy," i.e., the fact that lipid aggregates were not amorphous masses of round molecules but *ordered* structures of particles with heads and tails, resulted from physico-chemical investigations of, e.g., oil films on water surfaces.[73]

The bilayer got its epithets "Davson-Danielli," from a linchpin work of prewar membrane and surface chemistry that remained influential until around 1970, *The Permeability of Natural Membranes*, first published in 1943 by the British physiologist Hugh Davson and biochemist James F. Danielli.[74] Here, the protein fraction that was also found in membrane analyses was modeled as two extra sheaths adjacent to the double lipid film (see fig. 2). Davson and Danielli's bilayer should be understood in the context of different perspectives on subcellular structure at the time. In between the microscopic cell biological tradition and the molecular vision of structure endorsed by early X-ray crystallography, for example, the bilayer was a way of bridging morphology and biochemistry by looking at intermediate level structures. A similar avenue was tried out at the time by Joseph Needham's work, straddling chemistry, crystallography, and developmental biology, or in the work of German cell biologist Wilhelm Schmidt, who studied among other things the well-ordered molecular surfaces of frog retinae, using polarization microscopy as an indirect means to model the as yet inaccessible microworld of biomolecular structure and organization.[75] However, not only biological surfaces and interfaces, but also synthetics such as coatings and lacquers, were taken as exemplary of this world of neglected dimensions. It was not least their study, far and not so far from biology, that promised physiologists to get a grip on the vexing "permeability problem," i.e., the question of how living membranes managed to catalyze specific activities such as the uptake of a substance or the generation of electricity.

Let us now fast forward into the postwar molecular life sciences. Taking into account that knowledge about biological membranes resided almost entirely in indirect analyses, that is, nobody had ever managed to obtain an image of an intact membrane or studied it in isolation, the electron microscope must have appeared a boon to the field, particularly since the instrument rapidly transformed the image of cells.[76] Once methods of ultrathin section (microtomes) and fixation for biological materials had been established, glossy plates of micrographs from stained tissues or cells filled the pages of novel periodicals such as the *Journal of Cell Biology*. For membranes, as much as for other subcellular architecture, it may have felt as if the electron microscope stretched the realm of morphology and anatomy down to the molecular dimension.

Indeed, electron micrographs from the 1950s revealed two thin dark lines at cell boundaries separated by a lighter layer, sometimes termed the "railroad track model." These images were interpreted as suggesting a sandwich-like layer of proteins outside and lipids inside. This "unit membrane" was regarded as a confirmation of Davson and Danielli.[77] However, the issue was far from resolved. Electron microscopy required highly elaborate and harsh sample preparations. Staining procedures employing heavy metal compounds such as osmium or uranium were suspected to interfere with the structure of the delicate membrane layers, and subsequent drying of the sample led to shrinkage and consequently deformations. Thus, the procedures required for the imaging process, in which the actual biological material served more as a scaffold for a cover of heavy metals deflecting electrons than as the proper basis of the image, were prone to producing artifacts.

Moreover, different cultures of interpreting the electron micrographs' contours and shades existed.[78] Whereas researchers from Keith Porter and George Palade's department at Rockefeller University understood micrographs in connection with biochemical data from, for example, cell fractionation, others privileged EM as a direct visualization of ultrastructure, and rather tried to enhance resolution than to reconcile their data with other methods. The Swede Fritiof Sjöstrand maintained, in opposition to the Rockefeller group, that the membranes of mitochondria or retinal cells were not organized as continuous films of the unit membrane, but were formed by discrete, globular "subunits" shown in the bottom diagram of figure 2.[79]

Contradictory evidence and confusion regarding membrane structure surfaced in a 1965 Ciba Foundation symposium on *Principles of Biomolecular Organization*, which brought together structural biologists John D. Bernal, Francis Crick, and John Kendrew with electron microscopists such as Keith Porter and Walther Stoeckenius. Among viruses and proteins, membranes loomed large on the agenda. The extensive discussions on how to interpret electron micrographs, how to reconcile different specimens (EM had revealed a plethora of intracellular membranes), or how to take into account preparation methods nicely illustrate how the techniques of imaging complicated matters of structure rather than deciding them. Bernal, for one, derided sample preparation as "sophisticated cookery," depending not on principles, but on results.[80]

This is not to say that electron micrographs did not supply important information; however, the crux lay in their making, their interpretation, and the choice of the membrane specimen from amongst nerve, mitochondria, bacteria, etc.

Even worse for those interested in the physiological dynamics of membranes, EM could not resolve the "permeability problem," or how to explain membrane activities such as osmosis, substance transport, or the electrical excitation of nerves, since the instrument produced only static snapshots of a fixed specimen. George Palade jokingly asked his colleague Porter, on a trip to Berlin, to bring back an *Übermikroskop* showing sodium ions crossing a nerve membrane during an action potential. This remark illustrates not only what many considered a central problem of physiology emerging from Hodgkin and Huxley's model, but also that the ultimate goal was to make such "molecular events" visible, if only with an impossible instrument.[81]

In other words, a noncontroversial model of membrane structure in a living cell or tissue, or molecular mechanisms therein, as they were emerging in the 1960s for processes such as DNA transcription or translation, remained inconceivable on the basis of EM. The island of neglected dimensions was tucked in between the promise of life's next frontier on one side, and confusion or frustration on the other.

"The riddle of surface action"—membrane dynamics

As J. Loeb complained many years ago, obscure or inexplicable phenomena in biology are fashionably brought into the currency of "knowledge" by way of the philosopher's stone "a change in permeability." When to this the more modern elixir of "surface action" is added, night unto night sheweth knowledge. . . . But the sooner superficialities are replaced by a detailed understanding of underlying mechanisms, the better for science. We hope that this book will assist in defining what can, and what cannot, be done by the cell membrane, by "surface action" and by "changes in permeability."

H. Davson and J. F. Danielli, *The Permeability of Natural Membranes*, 1952[82]

In 1966, *Nature* featured a short note that reveals much about the persistent problems of membrane research. In fact, "Simple Allosteric Model for Membrane Pumps" was more a sketch of an idea than a paper describing experimental data. The contribution was authored by the American biochemist Oleg Jardetzky, later a pioneer in biological nuclear magnetic resonance spectroscopy (NMR).[83] It revolved around an illustration displaying how membranes, or more specifically the proteins sitting therein, would accomplish the transfer of substances across the impermeable lipid film (fig. 3).

In Jardetzky's model, the protein would form something like a pore in the membrane. The transport of substances across the membrane would be accomplished both by the docking and release of substrate from a binding site, as known from enzymes, and by the mechanical tilting of the protein subunits. These latter were depicted as rigid molecular structures spanning the membrane. Jardetzky's reference was the electrically active sodium/potassium ion "pump" of nerve, and his scheme

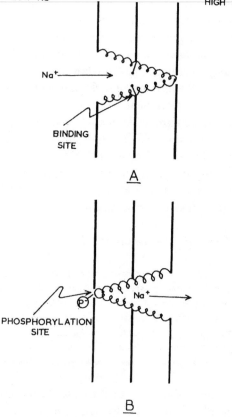

LEFT RIGHT

LOW Na⁺ HIGH Na⁺

Na⁺

BINDING
SITE

A

Na⁺

PHOSPHORYLATION
SITE

B

FIGURE 3 The "riddle of surface action" in the age of molecular biology. Diagrammatic model of how the nerve membrane could generate electric potentials, and more generally how substrates could be transported across the permeability barrier provided by a membrane. The three lines represent a cell membrane, the two spirals a membrane pore or transporter (possibly related to proteins), which binds the substrate (a sodium ion) on the intracellular side, then changes its conformation and releases the substrate on the other side. As many other metabolic processes, the reaction was thought to be driven by the addition of an energy-rich phosphate group (a process called "phosphorylation": see in glossary under ATP). The diagram implies a "pushing" or "pumping" of sodium out of the cell against a concentration gradient. This model drew both on the allosteric theory of enzyme regulation and studies of "pumps" in postwar nerve and muscle physiology. However, it was based on indirect data and conjecture, remaining speculative at the time, as indicated by the bare pencil-sketch style that leaves out many biochemical details. Since the 1990s, when related "alternating access" models of membrane protein function became widespread, this paper has been cited numerous times and is considered a conceptual origin. From Jardetzky, Oleg. 1966. "Simple Allosteric Model for Membrane Pumps." *Nature* 211, no. 5052: 969. Reproduced with permission.

proposed to understand its dynamics by a "molecular mechanism" analogous to macroscopic technologies: "It [i.e., the pump, M.G.] operates on the same principle as the locks in overland waterways, by altering the potential energy of the transported species."[84]

This model may serve as an entry point to understand the state of membrane transport research in the 1960s. The schematic character of Jardetzky's mechanism, which he stressed as one of its virtues, looks hopelessly simplified when compared to the sophisticated graphics of molecular mechanisms as presented in the Introduction. What was new about this model at the time, and what did it promise to explain?[85] The thrust of this paper was that its graphic and argument not only rendered Arthur Pardee's mysterious membrane activity visible, but also promised to understand it within the framework of 1960s molecular biology. Jardetzky mentioned for instance the allosteric theory of enzyme regulation as proposed by French molecular biologists François Jacob and Jacques Monod. Allostery formulated a model for how physiological activities could be regulated on the molecular level by the binding of effector molecules to enzymes. Not only did this approach straddle biochemistry and the cybernetic theory of signal transmission, but it depicted enzymes as objects that could undergo changes of shape, or transmit molecular movements in order to regulate a process.

If one reads Jardetzky's paper as the sketch of a molecular idea on how to address the lingering membrane problem, the nucleus of his idea was that physiological processes such as nerve excitation were to be explained by structure and dynamics of as yet hypothetical protein compounds (or complexes) within the membrane. To make membranes molecular, these dynamic, mechanically moving entities would have to be subjected to scrutiny by the cutting-edge concepts and techniques of molecular biology and biochemistry.

Membranes as black boxes

Jardetzky's proposal was certainly avant-garde pondering. Concerning experimental approaches to membrane dynamics, it would be an exaggeration to say that not much had happened since the interwar membrane craze, but it is fair to say that the problem lagged behind when it came to molecular explanations.

Davson and Danielli's 1943 book on *The Permeability of Natural Membranes* (second edition, 1952) had collated approaches and data on what researchers since the late nineteenth century, from botany to medicine, had referred to as the "permeability problem"—the vital question of how certain substances were able to cross the impermeable lipid films in living cells, whereas others remained outside. The problem was compounded when researchers realized that different cells effected specific "changes in permeability": Theories of "filters" or "sieves" in the membrane proliferated, and even a breakdown and reassembly of the membrane was discussed. Taking into account the electrical phenomena of the nerve membrane made things even more complex. Curiously, membranes of dead cells lost these inscrutable activities, and substances such as narcotics had an effect on them, making the problem appear as a hallmark to distinguish the living from the unenlivened.[86]

Davson and Danielli had differentiated cases of mere passive diffusion through membranes (acting like a customary filter) from those "where the laws of thermodynamics are apparently broken and molecules accumulate on one side of a membrane, in excess of the amount on the other side."[87] Such observations of concentration gradients actively maintained by living cells had been made on nerve, muscle, and other "excitable tissue," as well as on erythrocytes.[88] As Davson and Danielli considered a breach of the second law of thermodynamics improbable (which implies that concentration gradients tend to equilibrate over time), they inferred that the cells must supply energy: membrane activities, as part of metabolism, were a characteristic of life, without which it would end rapidly.

Insight into the coupling of membrane processes to cellular energy generation created a conjuncture of surface chemistry and membrane physiology with the burgeoning field of intermediary metabolism biochemistry. When radioisotopes of sodium or potassium (the ions central to the generation of action potentials in nerve) became available from cyclotrons in the 1930s, researchers followed their pathway through cells and studied their distribution across membranes, similar to how carbon isotopes were used in photosynthesis research.[89] Many other cross-connections between intermediary metabolism and the permeability problem existed, for example, Fritz Lipmann's characterization of adenosine-triphosphate (ATP) as the cell's central energy metabolite, driving all sorts of cellular processes, including membrane activities; however, transfers into membrane research (often carried out by physiologists working on cells or tissues) remained technically difficult. In essence, the central method of biochemists, to extract cell components and to analyze their properties, destroyed what was at the heart of membrane processes, namely the cell's architecture, and thereby the distinction between inside and outside.[90]

The most important membrane-related problem around midcentury was the generation and conductance of electricity by nerve.[91] Since

Helmholtz, this problem had kept scores of physiologists busy, but how nerve cells generated electricity across their membranes, leading to a potential difference that propagated along the fiber to carry an impulse, was a matter of controversy. In the 1930s, the squid giant axon became a model system of physiologists. As electrons could be directly inserted into this fortunately large structure in an approach called "voltage clamping," it became possible to display its electrical impulses on an oscilloscope screen as curves. The iconic spike, as seen also on paper slips, is found in biology or medicine textbooks ever since. After the war, the use of radioactive potassium and sodium tracers allowed physiologists to conceive of these impulses as an overlay of different, specific ion fluxes across the nerve cell membrane: The electrical impulse was generated by the decrease (depolarization) of the preexisting, negative resting potential—this could be recorded as an influx of positively charged sodium ions from the outside through the membrane into the nerve's cytoplasm. The rise of the membrane potential was then terminated by efflux of positively charged potassium ions—from the inside of the nerve cell into the medium, leading to a re-polarization of the membrane potential back to its original state. This, in a nutshell, was the model of the action potential published in the early 1950s by Alan L. Hodgkin and Andrew F. Huxley.[92]

Hodgkin and Huxley gained fame not only for detailing the sequence and the ion specificity of how the nerve membrane accomplished a process that was certainly as significant to biology as the copying of DNA, but also for modeling the membrane and its activity as a set of electrical elements—as resistors and condensers, or currents and charges that could be described mathematically. But even if this physical explanation of how electrical "signals" were generated and transmitted along the nerve was a spectacular feat of physiology, the explanatory scope of Hodgkin and Huxley remained on the level of the behavior of ions in bulk: It was about concentrations in- and outside, whereas what happened within the membrane, e.g., how the ions traveled in this specific sequence, and how a double layer of lipids and proteins orchestrated such a complex process, remained out of reach. Hodgkin and Huxley's action potential did not move down to the scale of "molecular events." This lack of a molecular dimension may have inspired Palade's fantasy of an *Übermikroskop* visualizing sodium ions crossing the membrane.

In other words, Hodgkin and Huxley conceptualized membrane processes in a living cell by macrophysical theories of reaction kinetics, electrochemistry, and thermodynamics, such as permeability coefficients, diffusion processes, or membrane potential equations. In analogy to physical theories, this approach can be called *phenomenological* as it reflected a statistical expression of how the measured electrical potentials and currents could be explained by the behavior of bulk ion fluxes. The membrane itself, represented by Hodgkin and Huxley as a resistor and condenser in a circuit diagram, remained a black box: a *microphysical* explanation of the molecular processes occurring within the membrane was wanting. Even if Hodgkin and Huxley concluded that "channels" of the membrane must be responsible for the in- and effluxes of specific ions in the nerve, saying what these were or how they functioned remained beyond the possibilities of their electrophysiological approach. Hodgkin and Huxley's "channels" of the 1950s resided on indirect inferences; however, the word was out.

In that sense, membrane physiology of the 1950s and 1960s had a scope very different to that of molecular biology. The model of semi-conservative DNA replication, for example, or that of protein synthesis by a ribosome moving along the RNA, focused exactly on the microphysical dimension of specific molecular processes in biology and rendered these visual. That is, the permeability problem went another way in the postwar decades than those of molecular biology[93]: A general "molecularization," which would have bound the different physiological, biochemical, and biophysical arenas addressing membrane activity into a common experimental, explanatory, and social framework was not seen, and theoretical proposals on paper such as Jardetzky's were only to mark this absence.

Pumps and transducers—metaphors in search of a substrate

If a historical line can be drawn from discussions such as Jardetzky's to the molecular "pumps" and "channels" inside the black box of the membrane, it may be worthwhile to briefly examine the origins of such mechanical concepts to describe life's dynamics. As these date back further than postwar mechanical models of genes and proteins, and bring to the fore physiology, this generally repositions their conceptual history.

The Cold Spring Harbor Symposium of 1940 was held under the heading "Permeability and the Nature of Cell Membranes," illustrating how important and somehow self-explanatory this problem must have been.[94] Physiologist H. Burr Steinbach discussed experiments on the ionic concentrations in cells and tissues, using isotopes, but also more conventional approaches. Steinbach was critical about the interwar models of the membrane as a sieve, or of selective permeability, as cells seemed to actively readjust ionic concentrations after perturbations—remember the connection between membranes and intermediary metabolism biochemistry. Pondering the fact that cells maintained a lower sodium concentration than their environments (just as in the abovementioned nerve case), he brought up "some mechanism of pumping" as an explanatory alternative to the sieve. Cautioning against the confusion of passive "permeability" and active "accumulation" of ions, his initiative suggested to "also take into account forces which actually do the moving."[95]

Note that Steinbach cautiously used the verb "pump" to address the active process, but remained agnostic about a possible entity behind it. Another physiologist at the time went further, speaking of "some sort of a pump," when discussing the "mechanism" of sodium excretion from cells.[96] Yet, also here, the question of what this pump was—the entire membrane of the tissue, parts of it, a component sitting in it, etc.—was not addressed, and even if usage of the term occurred in conjunction with the term "mechanism," the ways in which such assumed device worked remained elusive.[97]

In other words, whether the physiologists' "pumps" of c. 1940 were to remain terminological placeholders or metaphors (potentially productive, potentially dead) to describe a physiological process directed in space, i.e., the active transport of ions against a concentration gradient, was as open as it was for Hodgkin and Huxley's channels (as it turns out, the model for action potentials accepted today includes both—the channel for rapid passive flow of ions over the membrane, the pump for active exchange). The question whether there was a material correlate to these could not even be sensibly addressed in those studies, as they included neither biochemical work (such as preparations of membranes for chemical analyses), nor visualizations (such as by EM), but resided on the phenomenological level of bulk ion distributions across membranes. Physiology treated the membrane as a black box, and it was to keep doing so for decades, even if the 1950s saw some spatial models of how the transfer of substances across membranes could be accomplished by molecular entities.[98] However, most of these also remained agnostic regarding the molecular representation of this function.

A first step toward a materialization of the assumed membrane pump was the biochemical preparation of a membrane protein from crab leg nerves by the Danish physiologist Jens Christian Skou around 1960.[99] This mix of lipids and proteins was far from being a pure chemical substance such as an enzyme or DNA, but it was a *material correlate* of the membrane in the test tube, which allowed one to study its biochemical activities in isolation. Its activity was scrutinized by the methods of biochemistry, such as assays probing its activity by adding substrates. Skou observed a stimulation of ATP splitting by the membrane, when the potential substrates of transport, sodium and potassium, were added to the test tube. In other words, a positive correlation between ion concentration and energetic activity of the membrane preparation existed. In Hans-Jörg Rheinberger's terminology, these experiments substantiated the *conjuncture* between the physiologists' membrane in living cells and the biochemists' membrane as a material substance in vitro. One could also say that Skou's study moved the abovementioned different communities inhabiting membraneland—physiologists and biochemists, especially enzymologists studying intermediary metabolism, closer together. The fact that what I would call a first *materialization* of a membrane pump has been labeled as its proper "discovery," and that Skou shared a Nobel prize for it, underlines the relevance of these experiments in a development leading up to the molecular machinery that is the subject of this book.[100]

By the mid-1960s, the conjuncture of physiological and biochemical experimentation had corroborated that the characteristic sodium/potassium exchange in nerve membranes, which established the negative electric resting potential, was accomplished by an enzyme-like compound within the membrane. This latter could be characterized: It was known to be spatially oriented in the cell, its activity was stimulated by ATP (furnishing energy), as well as by its transport substrates potassium and sodium, and it was inhibited by poisons.

That is, membrane pumps had become more concrete since the 1940s, and this was part of the experimental basis on which Jardetzky sketched his model. However, even if the pumping activity (i.e., energy-dependent movement of sodium and potassium ions) could now be attributed to a biochemical membrane preparation, the alleged pumps had not been characterized as one or several molecular compound(s) of these preparations, which consisted of various proteins and lipids. Sometimes, the entire membranes, e.g., nerve, were designated as "pumps" and an accepted model of a process or mechanism on the molecular scale was still not available, even if the problem was on the mind of researchers.[101]

Thus, by the mid-1960s, membrane pumps had moved quite a bit from being hypothetical entities or metaphors without a material substrate; however, what this substrate was, and/or what mechanism accomplished the process, remained elusive and controversial. Just as membranes in general, pumps remained also far from becoming concrete molecular objects.

Receptors and transducers, or materializations of cellular communication in the cybernetic age

Meanwhile, other generic concepts to make sense of membrane dynamics took shape in 1960s sensory physiology, hormone, and vision research, that is, the sciences studying how the surfaces of cells and tissues interacted with their environments. Sensory perception was what had made Max Delbrück so enthusiastic about membranes, as he saw the possibility to push this field to the "limits of molecular biology." For what he and his coauthors broadly announced as the "next phase of our quest for a mechanistic understanding of life," Delbrück had again turned to a simple model organism, the fungus *Phycomyces*.[102] Rather than adopting electrophysiology to study nerve cells, which involved a plethora of processes in or around their membranes, he advocated to follow the successful example of how genetics had been molecularized by using simple phage instead of complex flies. Thus, he chose to study what he deemed a clearly delineated process in an easily studied organism: *Phycomyces*' growth away from light. Delbrück's aim in the *Phycomyces* project was a "transducer physiology" that was to single out the conversion of one environmental stimulus (light) to the first cellular "input" (in this case, cellular growth).[103]

Yet, the largely unsuccessful scrutiny of *Phycomyces* (which has in fact been called the "phage of perception," mistaking intention for effect) remained on the level of cybernetically framed behavioral studies, and thus a far cry from singling out the molecules of behavior.[104] Even after more than a decade of *Phycomyces* research, a molecular correlate of the "signal transducers" that Delbrück was so excited about in 1969 remained wanting. That is, precisely what he had demanded for molecular biology's next phase eluded him at the time—preparations of active sensory membranes or a molecular correlate of *Phycomyces* photoreactions.[105]

These specific shortcomings of his *Phycomyces* project reflected general problems of molecularizing membrane research at the time. The program of the 1965 Cold Spring Harbor Symposium on Quantitative Biology on "Sensory receptors" gives a similar impression.[106] The program straddled subjects as heterogeneous as hearing (the ear's mechanoreceptor hairs and the "cochlear transducer"), vision (eyes and retinae of diverse animals), taste and odor receptors, as well as bacterial chemotaxis. From many of the contributions, the aspiration to common models of "bioelectric transducers" can be gleaned. The broader direction of the meeting seems to have been to integrate electrophysiology, pharmacology, psychophysics, and information theory into a general cybernetic understanding of signal transmission and transduction depicted in an abstract way as circuits or switchboards. However, the lack of a coherent perspective on the subject is visible even on the terminological level: "Receptors," mostly used for hormone or chemical detection processes, were not clearly distinguished from, e.g., electrochemical "transducers." Moreover "receptor processes" were described

at very different levels—mostly on the physiological, ranging from entire organs down to their membranes, as studied by electrophysiology. A
proper molecular approach was outlined for vision only. Here, the cellular receptors of the retina (rods and cone cells being responsible for black/
white and color vision, respectively, converting optical stimuli into cellular signals) as well as the molecular receptor rhodopsin were known. This
photoactive pigment, sitting in the "disc membranes" of the retina's rod
cells, had been studied since the late nineteenth century, not least since biochemists had succeeded in preparing the reddish-purple substance in large
amounts from excised tissue (this fraction, called also "visual purple," took
a somewhat similar position to Skou's "pump fraction" from crab nerves).

Whereas visual physiology and psychophysics had detailed the reactions of cells or organisms to optical stimuli, biochemistry had, since the
1930s, characterized the molecular makeup and photoreactions of rhodopsin in the test tube. At Cold Spring Harbor, rhodopsin emerged as the
most diversely studied example of a molecular receptor: Ruth Hubbard,
Harvard professor, former collaborator and wife of physiologist George
Wald, discussed how "bleaching" of visual purple in the test tube (i.e.,
a color change of the material from purple to yellow upon illumination,
studied by optical spectroscopy) could be correlated to the process of
perception. The first "message" of the visual receptors to the nerve was
characterized by electrical measurements, and Hubbard pondered how
"conformational changes" of rhodopsin were correlated to the transduction of a light stimulus into an electrical signal of the nerve.[107] Referring
to conformational changes, that is, movements and changes of shape of
the rhodopsin protein, she adopted a key term of the unfolding mechanical discourse to understand protein function (see next section).

In that sense, vision was certainly something like a "pioneer sense"
for a combined physiological, biochemical, and biophysical study of the
receptor/transducer problem, and thereby membrane dynamics or "surface action." Yet, the precise relations of the phenomena as studied by
different techniques, let alone a molecular mechanism putting together
the activation by light, and the changes of protein structure and cellular
signal generation were not available.

This lack of a molecular approach to the receptor problem was even
more pronounced for other cases, when one participant at Cold Spring
Harbor picked up the old formula of a "change in permeability" to explain the membrane's electrogenic activity.[108] Time and again, the phenomenon was explained by rephrasing it, digging out the tautological
"philosopher's stone" that Davson and Danielli had hoped to bury decades ago. The molecular substrates and consequently the dynamics of

most receptors and transducers remained elusive, and even for an advanced case such as vision, research lagged far behind expectations and programmatics.

Proteins and the promise of molecular mechanisms

If Jardetzky's short paper leaned toward a molecularization of the permeability problem in the late 1960s, what he looked up to were clearly molecular and structural biology's successes, that is, the visual models of DNA and proteins, the recent molecular models of enzymatic regulation that went under the term of allostery, and the promise of a molecular biomedicine coming with it. Thus, molecular biology was the reference that many membranologists strived for, as it seemed to offer an ultimate level of physical explanation of life. Here, we shall briefly look at what protein structural biology had to say about molecular mechanisms in the 1960s. In fact, the elucidation and modeling of protein structures by X-ray crystallography flourished at the time, expanding its scope beyond the few molecules that had been tackled before. Moreover, as the method was becoming more articulate and less laborious, it was practiced beyond the field's early centers such as Cambridge or Caltech.[109]

The first edition of *The Structure and Action of Proteins* (1969), an introductory textbook coauthored by the architect and illustrator Irving Geis and Caltech's structural biologist Richard E. Dickerson, documents the breadth and vigor of structural biology's vision of how to explain life on the molecular scale. This book, which was to become a classic, promoted a simplified, visually appealing style of explaining molecular diversity and function. Photographs of complex, asymmetrical ball-and-stick models of molecules, for example, were replaced by redrawn schematics of molecular structure, or "outlines" of molecules. Moreover, cartoon-like sequences of such models were taken to display physiological processes (plate 2).[110]

Dickerson and Geis's images of life at the molecular scale resided on recent insight from protein sequencing (as pioneered by Frederick Sanger on insulin) and enzymology, but most of all on the modeling culture and language of structural biology.[111] However, the authors' narratives and imagery took the analogies between organisms, macromolecules, and machines to a further level, anticipating the molecular-mechanical vision of the coming decades. In their words, the living organism was a "complex factory," with small molecules such as vitamins serving as "the nuts, bolts and cogs that keep the wheels [. . .] turning."[112] Proteins were "the most remarkable chemical substances within the living organisms," playing "two distinct and separate roles: as structural materials and as machines that operate on the molecular level."[113] The main thrust of Dickerson and Geis was to explain how the structure and dynamics of these protein machines (NB: the generalizing expression) carried out various physiological processes.

Prime examples were the oxygen-binding proteins of blood and muscle, hemoglobin, and myoglobin. Hemoglobin had been a subject for protein studies in medical research, physiology, and biochemistry since the nineteenth century. It was rapidly obtained as a material substance in a pure state from blood and could be used to monitor physiologically significant changes (such as a color change upon binding of oxygen) in test tubes.[114] Due to this easily accessible materialization and its medical relevance, hemoglobin had advanced to a trailblazer for molecular and structural biology in the postwar years: Mutated hemoglobin genes were correlated to a functionally impaired protein that caused sickle cell anemia, the first "molecular disease." No less important were the structural models of hemoglobin established by X-ray crystallography (the protein crystallized spontaneously, a fortunate exception for structural biologists) from Max Perutz's group at the LMB, Cambridge, which spelled out important physiological phenomena as conformational changes of the protein.[115] To estimate the role of these models, one should keep in mind that by 1969, X-ray structures of only four (small and water-soluble) proteins were known.[116] For Dickerson and Geis, hemoglobin became the molecular "carrier" and myoglobin the "container" of oxygen in the body.

The science of proteins, embracing structural biology as well as sequencing and enzymology, promised to render all sorts of biological functions molecular—from hereditary, to metabolic, immunological, and ultimately cognitive: Basically, all of life's diverse processes appeared rooted in *interactions* or *conformational changes* of proteins. These processes were often described in the framework of classical mechanics, with proteins docking, binding substrates, moving in space, flipping around, rotating, being attached to one another, split, cleaved, etc. Depicted as cartoons, a space of "molecular action" within organisms emerged, in which proteins somewhat self-evidently began to figure as molecular machinery, and their substructures or atoms as moving parts.[117]

To understand the position of the novel molecular mechanical vision, one should mention that this was not the only fundamental and generalizing conception to understand life at an ultimate level. There were, for example, attempts at a "quantum biology," in which the electronic processes governing chemical bonds would be at the basis of physiological phenomena. Such a "submolecular biology" was advocated by muscle researcher and bioenergeticist Albert Szent-Györgyi or Alberte and Bernard Pullman in France.[118] Moreover, as discussed in the next section, bioenergeticist Peter Mitchell championed yet another understanding in which electrochemical processes as known from batteries or fuel cells would take the place of molecular mechanics.

In contrast to the membrane problem, some cases that took protein mechanisms beyond the more speculative level already existed at the time of Dickerson and Geis; for example, when structural biology merged with enzymology to connect spatial modeling of proteins with a chemical understanding of enzymatic reactions. At Cambridge's LMB, the groups of David Blow and Brian Hartley pioneered this merger by focusing on the digestive enzyme chymotrypsin, a protein cleaving proteins into peptide fragments. As for hemoglobin, chymotrypsin was easy to tackle, and even commercial preparations were available. These could be used for advanced techniques such as X-ray crystallography or protein sequencing, which required high amounts of a chemically pure substance.[119] The chymotrypsin work of the structural biologist Blow, experienced in Max Perutz's hemoglobin project, and Hartley, a biochemist in Fred Sanger's department, brought together the structural and enzymological approach to proteins step by step. In 1969, the team presented chymotrypsin's "mechanism of action" in *Nature*. That is, a combination of protein sequencing and structural biology had allowed them to map its enzymatic function onto its 3D-macromolecular structure.[120] "What happened" at the enzyme's catalytically active site in molecular terms was modeled as a sequence of chemical reactions in space, both as structural formulae and as ball-and-stick models.[121]

The molecular mechanisms and the machine analogy turned into far more than connotation as well when protein or nucleic acid components isolated from cells were used to "reconstitute" biochemical reactions; that is, when biological processes were staged from their component parts in the test tube. Studies on protein synthesis, which contributed to the elucidation of the genetic code, are a good example.[122] In the 1960s, the Indian-American chemist Har Gobind Khorana incubated synthetic nucleotides with cell fractions and enzymatic substrates to produce messenger RNA or polypeptides in vitro. This experiment not only helped to understand the biological protein synthesis as a set of chemical reactions, but also moved it into the realm of human control and intervention (see chapter 3).

Structural biology's expansion from a few model systems to the entire diversity of the subcellular microcosm also was an issue of scale. Dickerson and Geis, for example, take their readers on a veritable journey through the whirling microcosm not only of molecules, but also of cells, which culminates in a sort of "molecular drama" of five acts—the destruction of a bacterium by the immune system's proteins, ending in the demise of the invader after its membrane is permeated. The next step for structural biology would be to move up in order to understand, in the authors' words, "complex macromolecular systems" or "multienzyme machinery,"[123] that is, structures in the range of the neglected dimensions. Moving crystallography up from small molecules toward the cell's supramolecular organization, or, correspondingly, moving other techniques such as EM down to this level, was attempted within and beyond the hubs of structural and molecular biology. At Stockholm in 1964, for example, Munich biochemist Feodor Lynen, one of the few dynamic biochemists who worked on enzymatic reactions and metabolism in West Germany after World War II, had presented electron micrographs of the fatty acid synthetase complex, a large assembly of enzymes. The "giant particle" (*Riesenpartikel*), as one of his PhD students, Dieter Oesterhelt, had called it, emerged as a torus-shaped object (fig. 4).

This complex sat somewhere in between the cosmos of molecules and that of subcellular structures.[124] Membranes were another case in point for these neglected dimensions; and for their general relevance within physiology, they may be the most important one. However, if EM had produced conflicting data on membranes, X-ray crystallography was even less suitable to address this problem: Biochemistry's membranes, such as Skou's crab leg nerve preparations, were mixed and insofar impure aggregates of proteins and lipids. They could not be separated easily, let alone crystallized, which was a prerequisite of X-ray analyses such as for hemoglobin or DNA. Moreover, the precise assembly and function of proteins and lipids within the membrane remained unknown.

The membrane frontier

The molecular-mechanical vision of life as adumbrated by Dickerson and Geis, which was to spell out protein dynamics and move in scale and diversity over the entire range from cells to molecules, thus stands in stark contrast to the lagging advance of the membrane field.

This contrast, as well as the programmatic statements by, e.g., Delbrück or Arthur Pardee, may have acted as an attractor to take up the challenge.[125] This could explain why a number of young biochemists and biophysicists finishing their PhDs in the second half of the 1960s

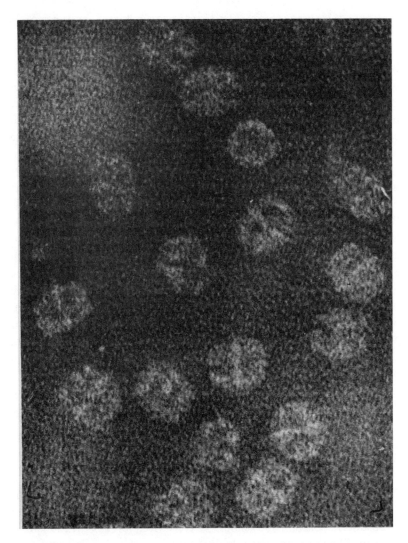

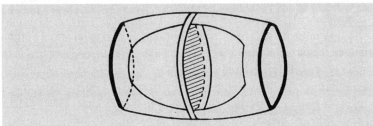

FIGURE 4 Approaching a world of neglected dimensions. Images of a large protein aggregate, the so-called fatty acid synthetase complex from yeast, in Dieter Oesterhelt's doctoral dissertation. Top: Electron micrographs showing contours of particles in between the dimensions of molecules and cells, taken in the lab of Peter Hofschneider, MPI of Biochemistry, München. Bottom: Schematic model of the molecular structure and assembly of the complex. These images seemed to suggest that there was a cosmos of structured molecular objects in cells that accomplished biochemical reactions in space, i.e., as processes that could be imagined to function like mechanical processes. From Oesterhelt, Dieter. 1967. *Zur Kenntnis der Fettsäuresynthetase aus Hefe*. München: Dissertationsschrift Ludwig-Maximilians-Universität München. Reproduced with permission.

were drawn to this difficult and, for many, uncommon subject. Remember that most previous membrane studies had been carried out by physiologists or physical chemists, whereas enzymologists or structural biologists had shunned the cell's boundaries as overly complex and intractable using their methods. Among those moving toward membranes was Dieter Oesterhelt, who had become interested in supramolecular structure in Feodor Lynen's lab when working on the fatty acid synthetase complex. He decided to take a sabbatical in the laboratory of the Rockefeller membrane electron microscopist, Walther Stoeckenius, who was to start off his own department in San Francisco in 1969. In the same years, Richard Henderson, who had contributed to the elucidation of chymotrypsin's "molecular mechanism" in Blow's LMB group, remembered getting the advice that soluble enzymes were "all basically solved" from a structural biology point of view, whereas membrane proteins were the new frontier.[126] Thus, he set out to do what must have looked like the take-home message from Jardetzky's paper (and yet was thought impossible by many)—to establish an X-ray structure of the "pump" that generated biology's most famous spike, the action potential. Henderson, Oesterhelt, and Stoeckenius (who represented an older generation) will be the three main protagonists of the following chapter who contributed, among many others, to turning around the problem of membrane dynamics, or, in other words, the decade-old "permeability problem" or "riddle of surface action."

Conclusion

Taking the different histories of membranes in structural biology, physiology, and biochemistry into perspective, the "pumps," "receptors," and "transducers" of the 1960s should be considered less as operationalized concepts like the gene, which had found material and molecular correlates since the DNA structure of 1953 (if also with multiple references). Rather, within a membrane science that remained one remove from providing molecular explanations, the molecular machinery described in this chapter remained a programmatic placeholder for ill-explained phenomena, often placed within a cybernetically framed discourse that promised to uncover general patterns of life related to membranes, such as chemical and electrical messaging, or cellular signaling and communication.[127]

A transformation of these pumps, receptors, and transducers, involving further, specific materializations and molecularizations comparable to the attribution of DNA, RNA, and proteins to genetic processes, would gain real momentum only in 1970s membrane research, driven by the upsurge of new objects of study, the transfer of biochemical and biophysical techniques, and a new membrane model. The hope of Delbrück and others to discover *one* general explanation of how membranes mediated signals between organisms and their environments, or how they transported substances, was to quickly evaporate in light of the diversity of molecules and mechanisms uncovered. However, a generalized molecular-mechanical vision of life, as outlined in Dickerson and Geis' *Structure and Action of Proteins*, or ideas on a "biomolecular organization" straddling molecules, aggregates, and cells, provided the background for a new generation of life scientists, e.g., at Cambridge's LMB or the MPI of Cell Chemistry at Munich, who worked on protein structure and function. The developments, topics, and people described in this chapter were certainly not the only roots to the membrane moment of the 1970s. However, they contributed significantly to moving life's machinery further down the road from placeholder toward materiality, and thereby effecting a change in how the science of our times thinks of and acts upon life.

2

Active Matter

Structure without life is dead. But Life without structure is un-seen.

John Cage, *Lecture on Nothing*, 1961, p. 113

In the 1970s, membrane research exploded. Where pre-viously there had reigned relative stagnation mixed with hope for a new molecular approach to biology, the con-crete results now piled up rapidly. Novel methods finally permitted membrane proteins to be isolated biochemically; spectroscopies provided insight into the molecular dynam-ics of membranes; and material models such as liposomes allowed membrane processes to be "reconstituted" in the test tube. All of these developments will be the subject of this chapter. A new general model of membrane organiza-tion, the "fluid mosaic," was rapidly accepted after 1972, and quickly put to rest debates about the 40-year-old Davson-Danielli structure (fig. 5).

Fast forward half a decade after this "membrane mo-ment" to the late 1970s, and biological processes from bioenergetics to neuro- or cell biology would be explained within a general framework of membrane proteins per-forming mechanical motions. Furthermore, in the 1980s, signal transduction, membrane protein structure, as well as membrane technologies became hot topics, all related to the premise that the membrane and protein dynamics would be spelled out as "molecular mechanisms."

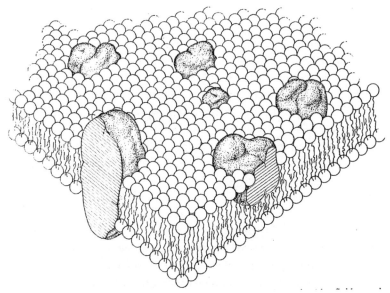

FIGURE 5 The dynamic topology of membranes. The "fluid mosaic" model of membrane structure according to Singer and Nicolson. Membranes are displayed as a matrix of tadpole-like lipid structures and proteins. Note the important differences to prior models depicted in figure 2: In the fluid mosaic, proteins are thought to represent asymmetric "lumps" sitting in or traversing the membrane (rather than being attached to it from the outside). Based on an integration of results from, e.g., immunology or biophysics (such as on the speed of molecular movements within the membrane), the fluid mosaic is thought to represent a "two-dimensional solution" of oriented proteins and lipids (Singer and Nicolson 1972, 720; Morange 2013). As much as Davson and Danielli's bilayer in the 1930s, versions of this figure (e.g., with a curved membrane sheath) became emblematic of the membrane moment around 1970, being reproduced countless times in textbooks etc., and the model is considered as valid ever since. To exaggerate: this model became the double helix of membrane research. From Singer, S. J., and Garth L. Nicolson. 1972. "The Fluid Mosaic Model of the Structure of Cell Membranes." *Science* 175, no. 4023: 723. Reproduced with permission.

Here, the surge in and transformation of membrane research will be told through the looking glass of the unfolding of research on a newly characterized research object, bacteriorhodopsin, a molecular pump similar to the examples discussed in the preceding chapter, which was isolated from active membranes of a bacterium. BR rapidly gained ground as a model system for the long expected molecular approach to the topic. This is not to say, however, that this was the single, most important story of the membrane moment around 1970—similar episodes could be told from the perspective of photosynthesis, neurophysiology, or cell biology.[128] However, zooming in on the history of one concrete object, and following it advance on the level of laboratory practice from a curiosity toward a globally researched model object, enables us to understand the role played by specific materializations of membranes and proteins in the transformation of this field, and the subsequent rise of a molecular mechanical style of explanation. As a tangible material substance, as stuff or *Stoff* endowed with a specific activity that indeed smote the senses of researchers, and that lent itself to various experimental approaches, BR has a trajectory that emblematizes a novel, integrated approach to the membrane problem. Due to its properties, it brought together the membranes of the physiologists, the (bio-)chemists, the biophysicists, and the electron microscopists, which existed disconnected before. Thus, what started as an encounter with a curiously colored lump of matter endowed with specific activities was to bring true the promise of membranes in the 1960s, if also in an unexpected way.

The new, simple model object BR rapidly garnered the attention of a wide range of researchers in "membraneland," from bioenergetics to physiology or structural biology.[129] This chapter will argue that material availability, accessibility, and activity made BR an object of choice for researchers in San Francisco, Munich, Cambridge, Moscow, and many other places. Among the people and places involved were major figures and renowned institutions of molecular biology, but notably also a younger generation of researchers who used this "material opportunity" to get a grip on the membrane problem, and who were to shape their fields for two to three decades after 1970. By relying on sources such as grant applications, laboratory notebooks, correspondence, and interviews, and by comparing them to the "folk history" of BR's discovery, this chapter will complexify the existing narrative, and highlight contingencies as much as the contributions of different actors.

In sum, this history of the membrane moment may also be read as a sketch characterizing a new beginning and a new generation of molecular life scientists after 1970, who were inspired by molecular biology's approaches and explanations, but also by a hope for medical and biotechnological progress that has been central to the life sciences ever since, not least in optogenetics, a contemporary method of the neurosciences employing exactly the molecular machinery the emergence of which is described here.

From membrane images to membranes as Stoff—*Rockefeller University, 1960s*

The Cytology Laboratory led by Keith Porter and George Palade at the Rockefeller Institute in New York City had been a hub for electron microscopical research on cellular ultrastructure since the 1950s. Methods were introduced, standards of how to interpret micrographs were being forged, and Rockefeller published a specific *Journal of Cell Biology* replete with glossy plates detailing mitochondria, membranes, etc.[130] Thus, it is no surprise that Walther Stoeckenius, a German medical researcher in his 30s, chose the Cytology Laboratory for a one-year Fulbright fellowship to the United States. Stoeckenius was a membrane microscopist: He had previously conducted both light- and electron microscopy to address the problem of structure, studying among other things so-called myelin figures, membranous aggregates that spontaneously formed when lipids were dissolved in water, and that resembled the myelin sheath of nerves. When Stoeckenius fled the academic "backwater" of Hamburg in West Germany and arrived in New York City in 1959, Porter advised him to continue to study membrane structure using electron microscopies of actual nerve myelin.[131] Stoeckenius, however, had also become involved in the debate about the arrangement of mitochondrial membranes, in which the Rockefeller Laboratory argued that electron micrographs of thin sections of cells and tissues had to be interpreted in conjunction with histological evidence from light microscopy and the biochemical composition and function of materials prepared from cells.

One may wonder why Stoeckenius reacted to a 1963 *Journal of Cell Biology* paper, which claimed that an obscure microbe, the salt-loving *Halobacterium salinarium*, might possess a "subunit membrane," that is, a membrane not made up of continuous lipid film as proposed by the Davson-Danielli or unit membrane models, but from spherical vesicles or "blobs" sitting next to each other (subunits were a competing membrane model depicted in figure 2). A subunit model was at odds with what was accepted at Rockefeller, but remember that membrane structure was still controversial at the time, as the question whether there was one structural model for all biological membranes, or different ones.[132] Moreover, Halobacteria were exotic objects of study—generally, Stoeckenius and the Rockefeller department focused on membrane specimen of medical or physiological relevance. By contrast, the curiously red-tinged microbe thriving in salt lakes such as the Dead Sea could not be claimed to have any such significance whatsoever; it had been studied by a few non-medical microbiologists in ecological contexts.[133] In short, hardly any molecular or cell biologist of the mid-1960s, let alone medical researchers, would have ever heard of the organism and whether what was true for Halobacteria was true for elephants must have appeared more than debatable.

Yet, there were better reasons than arguing about "subunits" or uncommon organisms that led Stoeckenius to pick up this topic. The *Journal of Cell Biology* paper included not only the glossy plates of cell sections typical of EM at the time, but also micrographs of biochemical preparations from the bug's membranes. That is, the authors showed images of the membrane as a *material substance* in the thick. Moreover, it was noted that these preparations could be obtained by simple experimental procedures, such as the transfer of cell materials into distilled water (Halobacteria live in concentrated brines). That is, Halobacteria seemed to possess a membrane that was both easily visualized in the electron microscope and was easily prepared biochemically.

In this respect, these specific membranes promised to be a "material opportunity": They *lent themselves* to combine electron microscopical with biochemical studies of membranes, which was not always feasible, and precisely the approach practiced at Rockefeller. In other words, these potentially idiosyncratic bug membranes promised to bring together the structural and the biochemical membranologists of the 1960s.[134]

Stoeckenius' grant applications reveal that the red bugs represented one option within his broader membrane project. In 1963, he submitted a proposal with Palade, headed "Fine Structure of Cellular Membranes"; within two years, he had hired a microbiologist, microbes were being grown at Rockefeller, and the membrane material was prepared and fractionated (i.e., separated into components with different properties) in order to distinguish components of the cell surface for visualization.[135] By the mid-1960s, Stoeckenius had hired a microbiologist, microbes were being grown at Rockefeller in fermenters, fractionated in distilled water, and the resulting material was centrifuged to further separate the cell surface substructures. As *Halobacterium* cultures possessed a striking red to purple hue, fractionation products differing in color—orange, red, and purple—became distinguishable and could be isolated in further rounds of centrifugation: Here was the elusive cell membrane as a material substance, as a *stuff* that one could see with the bare eye, and that could be literally sucked or scooped out of the centrifuge tubes.[136]

In essence, the advantage of working with the weird red bugs was that cell surface structures, visible in electron micrographs as dots or thin lines, could be matched with biochemical membranes. The striking colorations of these membrane substances, which contrasted to the mostly inconspicuous cellular preparations from other organisms, provided researchers with "indices in the test tube" to distinguish them, such as when one product of centrifugation was designated as the "purple pellet."

Moreover, these materials were plentiful, which meant that the chemical composition and enzymatic activities of the greasy, purple-reddish precipitates or the "supernatants" (i.e., what floats atop) could be tested and compared.

The endeavor to materialize membranes as preparations and to correlate their properties to images of the cell surface in situ continued in the second half of the 1960s. When a biochemically trained postdoc was assigned to further separate the different membranous components by using the ultracentrifuges in the department of Rockefeller cell biologist Christian de Duve, he fortuitously encountered a "purple band" in the test tube.[137] As this purple band (the term refers to its presence as a segment in the centrifuge tube) was shown to consist of membranous, sheet-like structures in the electron microscope, it was henceforth called the "purple membrane." This term is of interest since it highlights what scientists took this membrane to be at the crossroads of biochemical preparation and visualization: The name referred to *both* the biological structures in the microscopic images *and* to the fraction's most notable macroscopic characteristic.[138]

It is notable that the purple membrane was not even of central interest to Stoeckenius' project at this stage, but merely a collateral product on the way to something else. The purple membrane was even washed down the sink at the beginning as a possible contaminant or artifact of preparation, and thus rather represented something disturbing, even waste, than something worth questioning. Epistemologically speaking, at this early stage this materialized membrane had not properly acquired the status of an "epistemic thing" in the sense of Rheinberger, as it was not a focus of inquiry, let alone a "scientific object" in the sense of being well defined.[139] Even though the purple membrane resulted from sophisticated lab manipulations such as ultracentrifugation, it retained the unspecified materiality of a mass, a purple blob, or "something" at the bottom of the test tube with unclear reference to the membrane. The German term *Stoff* catches this vague, pre-epistemic status of the novel preparation very well: Stoff refers simultaneously to the macroscopic presence of material substances in the chemical sense (think of a jar filled with white powder) as well as to textile matter, that is something produced and structured, but still in bulk form and unshaped. The purple membrane was to remain mere Stoff for some years, as in 1967, Stoeckenius had accepted an assistant professorship at the University of California San Francisco (UCSF) Medical Center and his postdoc left the Rockefeller project soon after prematurely.

From Stoff *to molecule—San Francisco c. 1970*

A decisive turning point in a so far unspectacular project, as well as in the biographies of those involved, occurred when the problem of membrane structure in *Halobacterium* was taken up in a novel environment. In addition to generous funding offered to Stoeckenius at the UCSF Medical Center, he later remembered that the "attractions of California" had lured him to leave New York. This was also true for one of his early sabbatical visitors, Dieter Oesterhelt.[140] Whoever has strolled down Parnassus Avenue, toward the towering, white building of UCSF's Moffitt Hospital, nestled into one of San Francisco's hills, overlooked by greenery, and only a few steps from the Haight-Ashbury and Golden Gate Park, may not find this too surprising.

In addition to the marvels of land- and cityscape, UCSF was situated at the crossroads of the changing research topology of the life sciences in the Bay area. In 1967, UCSF Medical Center was not yet a hub of the "biotech" revolution that kicked in a few years later. On the contrary, the Cardiovascular Research Institute (CVRI), where Stoeckenius was hired, has been described as a site of "basic research" since the Sputnik shock.[141]

The Stoeckenius group at the CVRI focused on electron microscopic studies of biological membrane structure. Work on the colored membrane preparations resumed in 1969, when two young researchers arrived on sabbatical—the physicist Allen E. Blaurock and Dieter Oesterhelt. Oesterhelt had studied chemistry at Munich University and then worked as a PhD student in the laboratory of Nobel laureate biochemist Feodor Lynen at the Max-Planck-Institut für Zellchemie (fig. 6).

The prior focus of Oesterhelt's work had been predominantly enzymological, involving, for example, the characterization of reactive chemical groups within a large protein complex called the fatty acid synthetase. Yet, his 1967 dissertation also reveals an interest in macromolecular structure and function, a more general development in Lynen's department at the time (see chapter 1). He included a schematic model of the structure of the protein complex, as well as electron micrographic pictures that represented the *Riesenpartikel* (giant particle) as the torus-like supramolecular object shown in fig. 4, and mentioned the concept of a molecular radiating "arm" to explain the sequence of reactions carried out by the protein. While Oesterhelt had studied molecular structure before only in collaboration with other groups, he recalled that his interest in coming to San Francisco was to learn EM himself. Walther Stoeckenius appeared to him as one of the

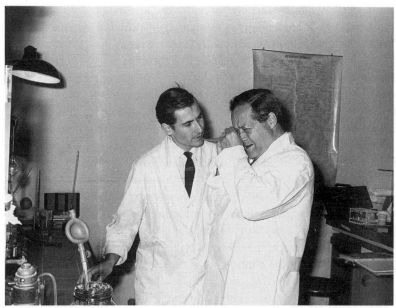

FIGURE 6 The excitement of molecular structure. Dieter Oesterhelt and his PhD advisor, biochemist and Nobel laureate Feodor Lynen, at the MPI of Cell Chemistry, Munich, 1967. Lynen is inspecting crystals of a large protein complex, the fatty acid synthetase, with a loupe. Crystals had been obtained in the course of Oesterhelt's PhD project, and they were considered a big step towards structural research on large active proteins. Copyright MPG/Krella. Reproduced with permission.

most renowned specialists, and he planned to leave his *Mittelbau* post (an intermediate level position in German academia) for a year to gain expertise.[142]

Stoeckenius had previously had visitors who complemented his electron microscopic approach with "wet" microbiology and biochemistry at Rockefeller, and one may assume that his interest in Oesterhelt was similar. The proposals submitted by the latter to the Deutsche Forschungsgemeinschaft (DFG, German Research Foundation) for a travel grant in 1969 provide further insight into where the project was originally intended to go. Oesterhelt mentioned EM, X-ray physics, and biological membranes as significant for his planned work on different macromolecular structures, such as ribosomes or enzyme complexes. The attached research plan described the halobacterial membrane preparations from Rockefeller, and mentioned the possible reassembly or the recombination of these preparations (and the microscopic fragments they contained) into a functional membrane.[143] That is, Oesterhelt wanted to learn about cutting-edge topics and high-tech instruments, whereas Stoeckenius required someone skilled in biochemical preparation, analysis, and possibly even synthesis, i.e., more customary laboratory handwork. As conceived at Rockefeller, the *Halobacterium* project offered both.

To reconstruct a detailed account of the experiments performed in the retrospectively decisive year 1969–1970 solely from the recollections of the three scientists involved is problematic, since the published accounts disagree on several issues, and since all accounts were conceived of years after these events that were to transform the whole endeavor and that had a significant impact on the actors' careers—in fact, both Oesterhelt and Stoeckenius would continue with the project started here throughout the 1980s and 1990s.[144] In this respect, Dieter Oesterhelt's laboratory notebooks have proven to be a very valuable resource, especially with regard to the impact of materiality on research. This neatly kept experimental diary is not a complete record of all the experiments that were conducted in the group. Yet, the notebook reveals how what appears post hoc to be a momentous "discovery" resulted from an extended and at times erring series of biochemical and structural biological experiments. Most notably, however, the experimental trajectory outlined in Oesterhelt's notebook shows the impact of the purple membrane's materiality on the development of the project. It becomes clear how the observations of and interactions with the purple membrane as active matter, i.e., as tangible "stuff that did something," turned the project from a general inquiry into membrane structure to that of a specifically active substance and molecule.

Oesterhelt's San Francisco notebook describes in detail 60 experiments carried out from September 1969 to July 1970. Each experiment has been numbered, and most of them dated. In addition to handwritten, comprehensive descriptions of the experimental set-ups, observations, results, and remarks, the notes also contain some printouts from instruments (such as spectrometers) as well as sketches, curves, and preparatory protocols. The electron micrographs mentioned in some of the experiments, however, must have been filed elsewhere.[145]

Upon his arrival at the Moffitt lab, Oesterhelt must have spent the first weeks of his stay growing *Halobacterium* and preparing the membrane fractions that had already been mentioned in the papers that had resulted from the work at Rockefeller. Even if preparations of materials for experiments are crucial for all biochemical projects, and often enough prerequisite as much as obstacle, their importance sometimes goes unmentioned in retrospective accounts, as, once established, the procedures become black-boxed as mere routine and they are often delegated to subordinate lab workers such as technicians.[146] Oesterhelt's records, however, show how a scientist unfamiliar with the culturing of a new organism, and the properties of its cells and biochemical fractions, appropriated step by step the established routines, reflected on unexpected observations, and tried out different approaches.[147] These careful trials are marked by frequent cross-checks with routine methods, and by extensive note-taking on minor issues, which went unmentioned later or in publications, including observations on the properties of cultures and preparations such as color or turbidity. One should understand these early, preliminary steps as both a form of comprehensive memory for further consultation, as well as a form of autodidactism.

Oesterhelt's primary goal was to reproduce the colored fractions as defined in a 1968 paper from Rockefeller, such as the purple pellet. One month into the project, he noted that indeed a round of ultracentrifugation *did* produce a "purple pellet, red supernatant," a result that he doubly underlined.[148] Furthermore, the indexical function of the preparations' materialities can be gleaned from the records: Oesterhelt sketched a centrifuge tube, he designated the fractions therein as "bright red" or "purple red," and as "without discernable structure, protein clouds," or composed of "big flakes in addition to small granula."[149] Such qualitative, perceptible characteristics of material substances, or *Stoffe*, must have guided Oesterhelt as he learned how to make preparations, and color was arguably the most important of these.

"Purple" and "red" became a shorthand for the respective material substances, and in mid-November 1969, Oesterhelt noted an effect that he would later highlight as a turning point of the project. Having separated "purple" and "red," he characterized these using optical spectroscopy and added acetone to the purple pigment. In a sort of aside at the bottom of the page, he added:

> Regarding purple pigment: Acetone causes immediate irreversible decoloration; yellowish 'protein' then precipitates with TCE [i.e., trichloroacetic acid], supernatant is colorless. Urea does not cause change of color, nor subsequent addition of dithionite.[150]

Even if instruments such as spectrophotometers or the electron microscope were used to analyze molecular composition and structure, big stretches of how this late twentieth century biochemical project has been conceived of and carried out at the bench appear to have been as much informed by Oesterhelt's chemist's skills and nose, that is, his sense of the perceivable properties and effects of tangible material substances.[151] Seemingly fuzzy categories such as flakiness, cloudiness, or color changes allowed Oesterhelt to orient himself in the maze of materials he encountered. So, manipulating materials with the hands and observing the effects directly did not lose their importance in the late twentieth century, when new technologies and instruments arrived. Not only did these skills allow Oesterhelt to understand and improve the steps of preparation, but they would be central to the most important observations he made.

The perceivable aspect of *Stoffe*, the relevance of which is highlighted by the notebook here, may be also described as their macroscopic or *molar* dimension. The term molar, from the Latin *moles*, a mass or bulk, can be used to designate aspects of a scientific object such as the body, or in this case matter,—the shape, color, immediate effects, or other tangible aspects—in opposition to its microscopic or *molecular* aspects, such as chemical composition. In a similar vein, Ursula Klein and Wolfgang Lefèvre have distinguished the perceivable dimensions of chemical objects, referring to visual appearance, etc., as opposed to models of imperceivable microstructure.[152]

Purple to yellow—an active membrane material

But back to the test tubes—what had happened here that apparently caught Oesterhelt's interest? The addition of well-known organic solvents, used to precipitate or separate substances, had caused the color of the purple membrane preparation in the test tube to change. When interviewed forty years later, Oesterhelt remembered the effect as striking and interesting to a chemist, and added that it set him on the track to determining what biochemical substance caused it, since he took the phenomenon as extraordinary among biological materials.[153] However, he pinpointed the observation to an experiment that was performed later, according to the notebook in February 1970, when he prepared samples of the purple membrane for analysis of their molecular structure by X-ray diffraction, carried out by his colleague Allen Blaurock. Then, as before, he added another solvent to the purple membrane in order to separate protein from lipids (it was known by then that the fraction contained both, as biological membranes did generally). With a yellow extract in hand, or so Oesterhelt said, he had asked Blaurock and Stoeckenius about the curious phenomenon. The 1968 Rockefeller paper had in fact mentioned in passing a similar color change of the purple material, but not much significance was attributed to the effect—the project being centered on the problem of membrane structure, and the colored fraction being considered more as waste than anything else. Oesterhelt, however, seems to have been alerted by what this material was able to do. Finally, a connection was made to similar color changes observed by Blaurock when conducting X-ray diffraction experiments on frog retinae in Maurice Wilkins' biophysics lab at King's College, London: Sometimes, the visual pigment of the retina, rhodopsin, also changed its color from purple to yellow (the protein was found in disc-like structures that had interested structural biologists in the postwar years).[154] In Oesterhelt's account, this analogy prompted him to go to the library, and search for chemical protocols to detect the chemical substance causing the color effect in the retina, which was known to be retinal (also called vitamin A aldehyde, the co-factor of rhodopsin). To demonstrate the presence of retinal in the purple membrane was to him the "discovery" of BR, as the protein behind the effect became baptized. The core of this story, formed by the analogy of the color changes between the two proteins, became widespread—Oesterhelt presented it thus, for example, at an award ceremony in 1999.[155]

Looking at the San Francisco notebook, the story looks more nuanced. It is true that the color change had repeatedly caught Oesterhelt's attention since November 1969, but it took until May 1970 for these observations to be addressed with an experimental strategy. A successful detection of retinal seems to have been carried out only in June 1970, and then within a few days shortly before his return to Munich.[156] In the meantime, Oesterhelt had conducted a series of biochemical experiments on the purple membrane—analytical ultracentrifugation, chromatography, or gel electrophoreses to separate its components, EM to obtain visuals of macromolecular assembly—all with the aim to characterize the material's components. He had tried various detergents (soap-like substances) to separate protein from lipid, the addition of some of which had equally caused the protein to bleach, i.e., to turn yellow, which was remarked upon repeatedly. However, the effect seems to have been discussed at first as a possible sign of *damage* to the membrane material— thus, more of a chemical artifact than a potentially interesting *effect* of the material.[157]

There are also other stories of how the conceptual stimulus to suspect retinal behind the color effect came about, all of which center on the spring of 1970.[158] Until July, when Oesterhelt returned to Munich, a few months if not weeks of condensed experimentation and hypothesizing must have taken place, which were to change the paths of both the project and its protagonists. Wherever the stimulus to think about retinal may have originated exactly, it is clear that Blaurock's prior experience with rhodopsin, in combination with Oesterhelt's attentiveness to chemical materiality, led to the transformation of this project. It is also clear, however, that this transformation occurred through a meandering path of experimentation that took several months until research had taken a new direction: The San Francisco project developed from an inquiry into the assembly of membrane structure to the investigation of a peculiar substance therein that changed its color under certain conditions. What had been waste at Rockefeller now became an object of research.

The relevance of specific skills and a sense for the significance of material behavior for this transformation can also be gleaned from retrospective accounts of similar projects at the time that did *not* take this new direction. The author of the 1963 *Journal of Cell Biology* paper that had originally inspired Stoeckenius at Rockefeller, recalled that he had also observed the purple fraction when in the 1960s—and kept washing it down the sink.[159] The substance had been present in a blatant sense, as tangible matter on the hands, but only a specific environment turned it into an object of research.

The chemistry of material activity

The conjecture that retinal caused the color change in the purple membrane coalesced in the last months of Oesterhelt's sojourn in San Francisco, when he adapted a whole body of chemical tests to analyze the composition of the material. This fueled the shift in focus from membrane to molecule.

By 1970, biochemistry and the physiology of vision were established fields and the active surface of the retina, what anatomists called its disc membranes, and the rhodopsin molecule were of interest to researchers worldwide, as seen at the 1965 Cold Spring Harbor Symposium on "Sensory Receptors," or in the 1967 physiology Nobel prize on chemical processes of vision.[160] Thus, besides the immediate interest that the retinal conjecture must have raised, as it seemed to reveal a biochemical link between evolutionarily distant organisms, there was a long-existing and reputable body of knowledge and technique that Oesterhelt could refer to in pursuit of this question.

Biochemical inquiries into purple-tinged retinal tissue had commenced in the late nineteenth century with German physiologists Franz Boll and Willy Kühne, the latter a successor to Helmholtz's chair at Heidelberg. From excised retinal specimens, Boll and Kühne had isolated a pigment, which they referred to as *Sehrot* or *Sehpurpur*, that is, visual red or purple. This substance displayed light-induced bleaching, a change from purple to yellow known from the entire retina. The finding that vision was somehow related to a photochemical reaction in the retina was analogized to photography, and had actually been used in the late nineteenth century to record so-called optograms, that is, transient images in the biological tissue, even of living animals.[161]

The term rhodopsin, referring to rose-red and vision in Greek, was introduced for the colored protein found in the retina by the American physiologist and biochemist George Wald.[162] Wald, one of the recipients of the 1967 vision Nobel prize, had taken up vision research in the 1920s. Following studies with physiologist and psychophysicist Selig Hecht at Columbia, Wald explored the chemistry of vision on a postdoctoral tour through renowned European biochemical laboratories, such as Otto Warburg's Kaiser Wilhelm Institute of Cell Physiology at Berlin-Dahlem.[163] When conducting spectral analyses of rhodopsin extracted from retinas, the conjecture of a carotenoid, that is, a vitamin-like substance, as the coloring agent of rhodopsin emerged. Retrospectively, Wald credited the Carr-Price-reaction (a test for carotenoids adapted from vitamin research, causing a bright blue coloration in test tubes) to have changed his life in the direction of a biochemical analysis of vision. The finding that the active substance of vision, which helped to convert light into a cellular signal, was related to vitamin A established a link to the burgeoning science of nutrition and vitamins. The relationships between foodstuffs and visual capacities in animals and humans had been researched since the nineteenth century, and around 1930 they coalesced in the knowledge of the chemical kindredness of biological substances from plants, vitamins, and visual pigments.[164] To make a long story short: Vitamin A, henceforth also called "retinal," was what made not only rhodopsin red, but also carrots, and this is the scientific rationale of why people were told that eating them enhances their vision.

Back to Oesterhelt and the purple membrane. The irony, or maybe the deeper message, of this project's development was that after having toiled with cutting-edge molecular biological techniques such as gel electrophoresis or EM for months, in summer 1970 he adopted the somewhat old-fashioned methods of retinal detection and studies of chemical bonds, thereby returning to what he knew best from home—natural product chemistry had been a mainstay of Lynen's lab in Munich.

He added the reagents for vitamin A-like chemicals known from vision research to the dissolved membrane and indeed found characteristic

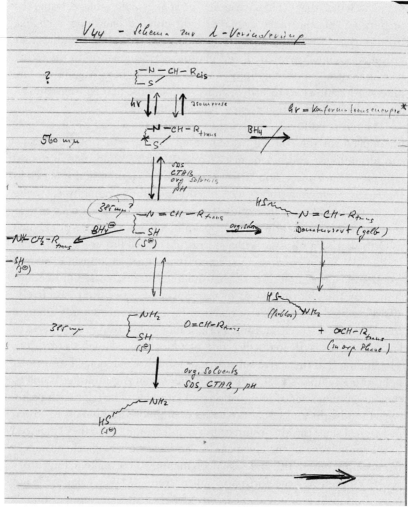

FIGURE 7 From membrane to molecule. Undated sheet from Dieter Oesterhelt's San Francisco lab notebook (presumably summer of 1970) displaying photochemical reactions of the purple membrane material upon the addition of reagents (noted next to the arrows, "SDS" [a detergent], "BH₄⁻" [a reagent known from organic chemistry], etc.). Oesterhelt noted the optical effects of these test reactions and used them as evidence of how the retinal cofactor may be bound to the protein (see formulae of chemical groups possibly involved in this bond, such as -NH₂, -SH, or -CH). Such studies of chemical bonds by reagents were a mainstay of organic chemistry, and thus illustrate the shift of the San Francisco project from analyzing membrane structure to exploring structure and action of a protein. From Oe SF, V44, MPG Archives (see note 145). Reproduced with permission of D. Oesterhelt, Martinsried.

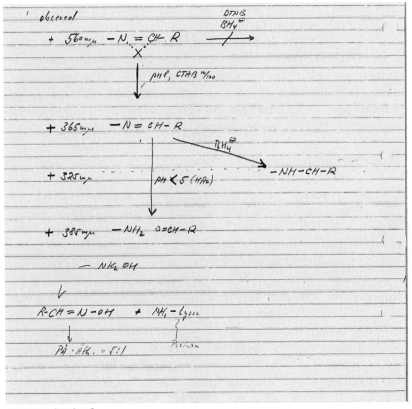

FIGURE 7 (*continued*)

colorations, among them the brilliant blue that had impressed Wald, which revealed retinal presence. A notebook sheet presumably dating to summer 1970, headed "Scheme of λ-changes," illustrates the impact of analytical biochemistry on the project (fig. 7).

The gist of this sketch is that Oesterhelt started to interpret the membrane's color changes in the test tubes as evidence for how the small retinal molecule was linked to the purple membrane's protein and which chemical reactions were behind its color effects. These direct, material indices of retinal presence from organic chemistry were corroborated by quantitative data from mass spectrometry, a contemporary physical technique to directly determine the weight as well as the composition of molecules. That is, both old and new methods were recruited in tandem to tackle the problem of the molecule under scrutiny—a research strategy repeatedly observed in this story.[165]

By summer 1970, the project deviated quite a bit from the original plans of Oesterhelt and Stoeckenius, as documented in their research proposal a year earlier: The focus was now the protein of the purple membrane, its similarities to the visual receptor rhodopsin, and the observed color changes, which must have looked utterly similar to what happened in the retina. Had the "molecular eyes" of bacteria been found?

Membrane structure rendered tangible

Yet, there is another line of how the San Francisco project developed that needs to be threaded in here, as it connected the biochemical studies of the membrane's molecules to analyses of their structure. Zoom back in time to January 1970. The purple-yellow effect observed by Oesterhelt in a test tube had probably not even given rise to a defined question, when another important feature of membrane materiality surfaced in the San Francisco notebook. Reporting structural studies of the purple material by X-ray diffraction, Oesterhelt noted the observation of a "hexagonal pattern of membrane."[166] In other words: Photographic records of membrane samples subjected to X-rays displayed a characteristic molecular arrangement. The membrane seemed to be composed of molecular structures ordered in regular hexagons, and this was quite remarkable, since such type of symmetry was only known from crystals. Was the purple membrane a crystalline structure?

Structural studies of the purple membrane were among the reasons Oesterhelt had actually come to San Francisco. To this purpose, he employed X-ray diffraction, a technique to characterize supramolecular organization, that is, shape and arrangement of molecules in the "world of neglected dimensions"—those larger than most molecules, but smaller than cellular components. The advantage in comparison to X-ray crystallography, the technique's "big sister" widely known from the postwar work on DNA and proteins, was that X-ray diffraction worked for all sorts of colloidal aggregates or subcellular structures that showed some degree of order, such as found in silk or cellulose fibers, in soap or detergent micellae, membrane stacks of chloroplasts, retinal rod cells, or the myelin sheath of nerve.[167]

What is more, such structural analyses of membranes had been the domain of Stoeckenius' other sabbatical visitor, the physicist Allen E. Blaurock. During his PhD, Blaurock had analyzed the rhodopsin-containing disc membranes of frog's retina by X-ray diffraction, which made him aware of their light-dependent color change. On a postdoctoral stint at the Medical Research Council's Biophysics Unit in London, he had continued along these lines with structural biologist Maurice Wilkins, among Watson and Crick the lesser-known third party of the 1962 Nobel prize for DNA structure.[168]

Just like Oesterhelt, Blaurock also remembered that he had unexpectedly encountered a structural pattern on a photographic plate that indicated crystallinity when exposing a suspension of the purple membrane to X-rays.[169] To check whether the observed pattern was caused by a contamination or by the membrane, Blaurock dried the purple membrane preparation on a surface, a procedure adapted from Wilkins' lab. Here, the materiality of the purple membrane again became important: By spreading the "membrane lump" on a surface, Blaurock *reconstituted* the original character of the material as a thin film. To put it differently, he created a laboratory model of a membrane that shared many properties with the molecular layer surrounding cells.[170]

Such reconstituted purple membrane films were well suited for structural studies: Whereas other laboratory models of membranes, such as myelin or erythrocytes, disintegrated when vacuum-dried on surfaces for experimentation, all this could be done to the purple membrane film without problems; moreover, the material was available in ample amounts that allowed for many trials.[171]

With these membrane films in hand Blaurock confirmed that they contained molecules in a crystal lattice. The question now arose whether the molecular organization could be attributed to one of the two molecular components in the fraction. Since the Rockefeller work, these were known to be protein and lipid. And here is the nexus point of Blaurock's structural story to Oesterhelt's analytic-chemical story told previously: It was for these attempts to separate the material's components that Oesterhelt added lipid solvents such as chloroform or ether to them, and this was how he stumbled once again on the color change. The conjuncture of these two lines of research could also explain why Oesterhelt remembered having observed the effect when preparing samples for Blaurock in an interview, even if it was on record before these experiments—the phenomenon may have only surfaced on his mind when it was discussed.[172]

After separating the two compounds, it was soon found that the hexagonal pattern was caused by the purple membrane's protein. In the ensuing X-ray and electron microscopic studies scrutinizing the structure further, the pattern became visible in a peculiarly direct way: When Oesterhelt analyzed a film specimen as prepared by Blaurock using EM, he noted a series of peculiarly oriented lines in the image. It seems that

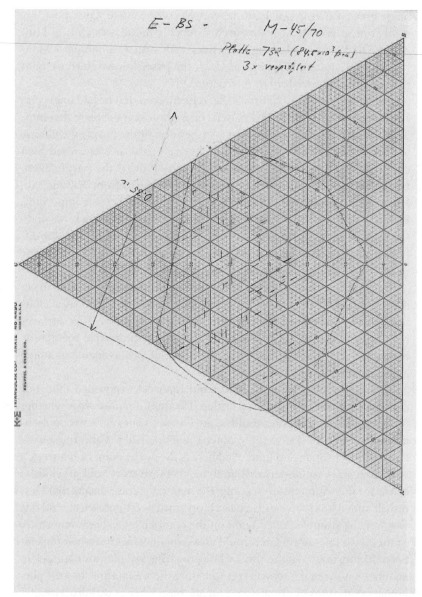

FIGURE 8 Molecular organization made tangible. Undated sheet of triangular scale paper inserted into Oesterhelt's San Francisco notebook. Numbers indicate the electron microscope run; outline and dimension of purple membrane patch are indicated by pencil. The material's break lines upon desiccation (similar to cracks in the mud surface of a dried-out puddle, indicated as short pencil lines), match with the 60° angles (and their multiples) of the triangular raster. Oesterhelt interpreted this as revealing the membranes' molecular organization, formed by aggregates arranged in a hexagonal symmetry (see figure 12). This "material insight" matched with data from X-ray diffraction. From Oe SF, filed following V27, MPG Archives (see note 145). Reproduced with permission of D. Oesterhelt, Martinsried.

Oesterhelt had conducted the structural analysis himself in this case, as the notebook contains preparatory protocols and descriptions of the results, even if the micrographs themselves must have been filed elsewhere. When interviewed, he remembered that the regularity of the lines' orientation appeared to him when plotting their patterns on triangular scale paper. In fact, a diagram in the notebook nicely shows how a structural pattern, displaying 60° angles between the lines, and thus some sort of hexagonal structures, became visible when the micrographs were transferred to the paper. Some weeks previously, Oesterhelt had already noted on a similar experiment: "Cracks of the membrane: precisely 60° [angles], that is, possibly caused by hexagonal structure (fig. 8)."[173]

When Oesterhelt had performed these experiments in January and February of 1970 and reflected on the regular cracks, Blaurock's observations of structural patterns indicating crystallinity were probably already around.[174] Yet, the causal connection between the structural pattern and the protein probably emerged through their collaboration, and the latter's plotting of the structure on triangular scale paper rendered the observations tangible. Again, it was communication between individuals from different research fields, and here in addition a specific way of representation that turned experimental effects into remembered findings.

Whatever the exact sequence of events, it had become clear through Blaurock's and Oesterhelt's experiments that the purple membrane was indeed composed of a protein in a regular, crystal-like arrangement—this was quite an uncommon feature of a biological membrane. More importantly, this turned the membrane into a material opportunity for structural biologists attempting to tackle the lingering membrane problem: Here was a membrane material that could be easily produced in large amounts and that displayed molecular order. Thus, analysis of a membrane by X-ray crystallography, as successfully performed for DNA and proteins, must have appeared in reach, with the *Stoff* resisting, not least, the harsh experimental conditions. The strange red bug's membrane, that is, tough and structured, was biological matter made to work with.

The new biology of membranes

In February 1971, the trio submitted two manuscripts to *Nature*. The first, authored by Oesterhelt and Stoeckenius, was a biochemical argument for the presence of a retinal-containing protein in the purple membrane. The structural data were separated and authored by Blaurock and Stoeckenius.[175] By the time the revised manuscripts went into print in early fall, the team had already ceased to exist for a year. Oesterhelt had returned to his post at the university's Institute of Biochemistry, in the Karlstraße of downtown Munich, while Blaurock had signed again to Drury Lane, London, i.e., Wilkins' Biophysics Department at King's College.

The papers had been relegated from *Nature* to *Nature New Biology*, a newly founded satellite of the classic journal, which published a number of important contributions on membranes during its brief existence.[176] Oesterhelt and Stoeckenius' contribution, entitled "Rhodopsin-like Protein from the Purple Membrane of *Halobacterium halobium*," listed spectroscopic, chromatographic, and chemical evidence about the size of the protein, its alleged co-factor retinal, and its binding to the protein. The novel naming of the molecule, however, only appeared somewhat couched in the last sentence, usually a site of speculations: "On the basis of these observations, we suggest that the purple membrane may function as a photoreceptor and propose the name bacteriorhodopsin for the purple membrane protein."[177]

This reluctance may be comprehensible in light of a referee's critique. Taking very much a biologist's perspective, the referee had doubted the analogy between the purple membrane protein and visual receptor rhodopsin on the basis of chemical analyses, and pointed to the evolutionary distance between the red bacteria and any organisms known to harbor rhodopsins.[178] This skepticism illustrates the unexpectedness and novelty of the findings from San Francisco, and their implicit message that a weird red bug may have photoreceptors on a molecular scale that were similar to eyes of animals.[179] Keep in mind that the paper indeed contained no data on what must have been the most central issue for a biologist—the function of this structure in the organism.

In a letter to Oesterhelt, Stoeckenius stated that the referee's comments were "clearly those of an old man, whose identity is not difficult to guess," thereby referring to George Wald, the most senior researcher on the biochemistry and physiology of vision. In fact, Wald may have been another reason why the paper also included the customary methods from rhodopsin detection in addition to modern data from mass spectrometry and gel electrophoresis.[180] The role of these "old" methods may have been twofold: Oesterhelt adapted them to rapidly detect retinal after the conjecture arose in summer 1970, while, in the publication, they were supposed to convince possible referees from a different fields and/or a different generation.

Blaurock and Stoeckenius' paper, published back-to-back, commenced on the striking X-ray diffraction patterns. The membrane's hexagonal lattice and the protein's regular arrangement were not only inferred from these patterns, however, but the direct visual evidence of the cracks played an important role in their argument as well.[181] The paper outlined a spatial model of the purple membrane's molecular structure as an ordered, two-dimensional "protein landscape."

Together, the twin papers from San Francisco, with a few following suit until 1975, established the purple membrane and BR as objects of science and set the stage for an entire research field. Publication activity rose exponentially after 1971—these papers had been cited more than 1,400 times by 2014—and laid the foundation for Oesterhelt's and Stoeckenius' further professional developments insofar as the laboratories of both would focus on BR and other membrane proteins studied by similar methods. It is certainly not an exaggeration to state that, among scientists, the 1971 short publications were what scientists worldwide would think of when hearing the name Oesterhelt or Stoeckenius.

However, the pivotal role of these papers and the way they present the "discovery" of BR obfuscate that their data and argument have resulted from a major shift of the project from a study of membrane structure toward the analysis of an active material substance prepared from it. This shift was the outcome of almost a year of meandering experiments, which unfolded on the basis of the prior experiences and skills of the three people that worked together at San Francisco. Most notably, a comparison of the archival records studied here with publications and interviews could not single out the one "eureka moment," but rather dissolve it into a chain of events that show coherency as a pathway toward a goal only when reading them against the papers. Thus, the momentous finding memorized by the protagonists may, to take up Frederic Holmes' interpretation of another momentous experiment from molecular biology, refer not to anything on record, but to the inner life of the participants. That is, as scientists, they were expecting a moment of discovery and striving toward it, and this experience may have even become real at one point of time.[182] However, this experience may not have been the same in all minds, it may not necessarily concur with the first instance of an observation on record, and it may not even be possible retrospectively to attribute it to *one* experiment isolated from the environment of the research group. Moreover, countless re-actualizations of this moment in light of later developments may have modified its memory. Regarding the differing accounts of Oesterhelt and Stoeckenius, for example, one should not forget that they became competitors shortly after 1973, leading to a complicated relationship.[183]

Whatever each individual may have contributed, the new membrane biology of BR has ever since been tied also a specific place, San Francisco. Thereby, I mean not only to Stoeckenius' Moffitt Hospital department at

UCSF: When Oesterhelt gave a talk at a 1972 biochemical meeting on a subject that was probably unknown to everybody in the mostly German audience, he started with a slide showing an aerial view of the salt works of San Francisco Bay, where square miles of red brine indicated the mass growth of *Halobacterium*.[184] In numerous talks, articles, and books that have followed, as well as in the conversations the author has held with members of the BR community, the fact that the purple organism, which had helped push membranes into the molecular age, thrived here has been evoked. When flying into the South Bay airport, or so the stories go, a visitor could discern the tinge of the blooming brines from above. A new biology had found its place, its time, and its narrative, and one that would generate further resonances between the protagonists and their Californian environment—such as the connections to NASA's nearby research facility, or the stories of biotech that were to unfold here soon after (chapter 4). In the fall of 1971, however, this was all but evident. One could also have considered the twin papers as unexpected data on an odd topic, or as merely chemical evidence on a substance with debatable biological relevance.

Nature's pleasant clue on membranes

The purple membrane had already made a splash before the *Nature New Biology* papers were out. Walther Stoeckenius had phoned Max Delbrück in 1970, traveled to Pasadena for a talk, and sent the manuscripts over after submission. Such "briefings" of colleagues—Delbrück still seems to have been an important figure in molecular biologies, with membranes and sensory perception ranging high on his agenda—can be understood as a strategy used by scientists to raise attention to their work and/or to acquire feedback from potential reviewers. In the present case, Delbrück thanked Stoeckenius for "exceedingly interesting papers on the subject of the weird purple membrane," said he was puzzled by the fact that the structural results were more advanced than those from the long-studied rhodopsin, but also showed some caution about the data.[185]

Next, Stoeckenius presented at the 1971 San Francisco meeting of the American Society of Biochemists a week before publication, *Nature* announced the findings in the main journal, talking about a possible "photoreceptor," and shortly after, Blaurock even mused in the *New Scientist*, a British science magazine, about the "purple eye of a bacterium" that may help to understand vision.[186] That is, the journalistic coverage on the results reinforced the impression, only implicit in the twin papers, that a simple system for a molecular study of vision had been found. In fact, Delbrück, and potentially others, may have thought for a moment that *Halobacterium* and its purple membrane was the long-sought "phage of perception" that would molecularize sensory and receptor physiology.[187] It was not long until he obtained a bacterial culture from Stoeckenius and assigned one of his graduate students, Lily Jan, to work on the topic at Caltech. Her first thesis committee report, probably from fall 1972, expresses the perceived promise of the new subject, when she aptly called the purple membrane a "pleasant surprise and clue offered by Nature."[188] However, Jan, following to a great extent molecular biological thinking, wanted to address the problem not with biochemistry and biophysics, but through genetics, i.e., by looking for dysfunctional mutants of the purple membrane. As with *Phycomyces*, the approach from Delbrück's lab did not have much impact, however, and in the early 1970s membrane research and molecular genetics remained disconnected fields (see chapter 3).

However, for many others, BR and the purple membrane indeed became the material crystallization nucleus to molecularize membranes, right at the time when the field was undergoing a general transition for which many of its protagonists from the 1960s had yearned for so long. The fact that the literal analogy to vision soon fell apart did not hamper the ascent of what was to become a model system for spelling out the molecular mechanisms of membrane proteins. In the remainder of this chapter, I will follow this development along the growing group of BR researchers until around 1975. Then, biochemical analyses of the new protein, the electron microscopic and other structural studies of the membrane, as well as cell physiology were bound together in a fairly coherent framework, and the concept of BR as a "molecular pump" had taken a clear shape.

Mechanical matter—Munich, 1970–1974

In summer 1970, shortly after the decisive work in San Francisco, Oesterhelt returned to Munich. He held the post of a *Konservator* at the university's Biochemical Institute, which was housed in the same building as Lynen's Max-Planck-Institute für Zellchemie. Since Lynen was also a professor at the university, Oesterhelt somehow returned to his orbit; however, he had taken the new topic home and collaborated further with Stoeckenius.[189] His laboratory notes start with a new "Versuch 1" (experiment 1) in September, that is, before the *New Biology* papers had even been submitted.

Oesterhelt largely continued with the biochemical analyses of the purple membrane protein, both on the level of wet chemistry and analyses by physical instruments (e.g., automated amino acid analyzers)—expensive equipment that belonged to Lynen's laboratory. A grant proposal submitted by Oesterhelt to the German Research Foundation in spring 1971 confirms that his interests and work style had remained largely within the instrumental and conceptual framework of enzymology and intermediary metabolism biochemistry.[190] In order to gather materials for these experiments, he tried out different preparatory procedures to separate and extract membrane components in 1970 and 1971, involving solvents, reagents, etc. These analytic labors were not always successful— Oesterhelt noted the "catastrophic results" of a harsh chemical procedure to isolate the retinal co-factor of the protein, involving ether as a solvent.[191] The failed experiment, which probably cost no small amount of time and effort, illustrates the difficulties of treating delicate biological matter with organic chemical methods. It also highlights that Oesterhelt had not quite turned into a molecular biologist following his year in San Francisco. By contrast, the new topic was treated in Munich very much by the customary organic chemical methods of Lynen's lab. Retrospectively, it even seems as if the experimental development toward the next milestone of this project was fueled by the preparatory, wet work with the purple membrane in the test tube.

From his correspondence with Stoeckenius at the time, it is evident that the trials with different solvents were meant to accomplish two things: Not only did Oesterhelt try to separate components of the protein in a better way, but he was looking for conditions under which a reversible color change could be obtained; that is, conditions under which the purple substance would turn yellow and then return to the initial state. In an interview, he described this as "functional thinking" inspired by enzymology: All enzyme-catalyzed reactions are reversible, and thus a reversible color effect could be taken as an indication that this was not an artifact, but a biologically functional process.[192] Note that Oesterhelt again approached the problem of biological function here very much from the standpoint of a (bio)chemist—by contrast, a molecular geneticist such as Lily Jan from Delbrück's lab looked for dysfunctional mutants.

At the end of February 1971, Stoeckenius, who had also embarked on similar trials of adding reagents and checking effects, wrote in a letter to Munich that when concentrated salt solution was added to the ether mix used to separate lipid from protein, the "material is reversibly bleached by strong white light! . . . We have repeated this game quite a few times and it works perfectly."[193]

Tinkering with the biochemical material and reagents, with the goal of isolation or that of obtaining a reversible process, resulted in a test tube assay that must have appeared like a "game" played with the membrane substance (the notion appears several times in the notebooks and correspondence): Shine a light, the purple stuff bleaches to yellow; return to the dark, it becomes purple again. In fact, I have set up the described assay myself, by adding sodium chloride and ether to a purple membrane preparation, and without any longer trials observed the bleaching. On a bright spring day, walking to the window was sufficient to obtain an effect striking enough to gather a lab crowd to marvel and discuss it (plate 3).

However, the "salt-ether system," as the assay was baptized, became much more than a game. It was a first step toward modeling material activity as a biochemical reaction of the purple membrane in vitro, in analogy to the known bleaching of the retina's visual purple. This assay, a simple and almost crude set-up to display material activity, epitomized a central theme of the nascent field: The purple membrane was a reactive material in the most tangible sense: Light on, yellow; light off, purple, etc., ad infinitum. It was "stuff that did something" or mechanical matter.

From color change to molecular mechanism — optical spectrometry

Obviously, the material's photochemical effect lent itself to being scrutinized by instruments. In response to one of the talks Oesterhelt held in Germany and the US, he was invited in March 1972 by Benno Hess, director at the Max-Planck-Institut für Ernährungsphysiologie (MPI of Nutritional Physiology), to give a seminar at Dortmund.[194]

Hess had been working with bioenergeticist Britton Chance at the University of Pennsylvania on topics such as "metabolic control," i.e., the regulation of glycolysis in yeast or tumor cells. Chance and Hess were not so much doing analytic or intermediary metabolism biochemistry as Lynen did, but molecular biophysics. "Pathways weren't my interest, mechanisms were," or so Chance stated in a retrospective interview — that is, he studied the molecular details of enzyme action with physical instruments.[195] Chance's department at Philadelphia, the Johnson Foundation, was renowned for the development of sophisticated mechanical and optical technology to study biochemical reactions, such as the "stopped-flow apparatus." By rapidly mixing enzymes and substrates with motor-driven syringes, it allowed metabolic reactions to be resolved at a high temporal resolution.[196] However, even the resolution of the stopped flow apparatus (which became a classic of biochemistry) was limited by the lag times and inhomogeneities of the mechanical mixing process. As pre-war studies to follow an enzymatic reaction by measuring fluorescence changes of the protein in a test tube by cell physiologist Otto Warburg had indicated, light was the means for an even more precise and continuous scrutiny of biochemical kinetics.[197] Elaborating this biophysical approach, Chance's department had developed the "double-beam spectrophotometer," which allowed synchronous illumination of a sample and recording of the emitted fluorescence.

Given what we know of the purple membrane so far, it is clear why Hess became interested in Oesterhelt's curious new topic—for spectroscopists and biophysicists, an optical effect that could be followed by the bare eye must have truly appeared as a "clue offered by nature" to study protein and membrane dynamics (to quote Lily Jan, see above). Moreover, it was an opportunity to profit from the double-beam spectrophotometer in Hess' department. For Oesterhelt, the collaboration was a chance not only to gather data, but also to introduce his uncommon subject to the German research community.

He traveled to Dortmund and suggested bringing along some samples of cells and membranes. To his host's query regarding how much of the auspicious material would be available, he replied that he could provide at will and proposed a complete scheme of spectroscopic experiments he wished to conduct.[198] The collaboration between Hess and the newcomer Oesterhelt must have thrived rapidly, as their frequent correspondence from 1972 to 1975 documents, with Hess offering Oesterhelt a position at Dortmund soon after. Their first joint paper in 1973 presented the kinetics of the purple membrane's photoreactions in the salt-ether-system, termed "bleaching and regeneration" in analogy to visual rhodopsin, as analyzed by the cutting-edge spectrometer coupled to a minicomputer.[199] Oesterhelt and Hess also noted that the optical effects (i.e., color changes as changes of absorption or emission maxima of light waves) were accompanied by a chemical change, that is, a de- and re-protonation of the substance in the cuvettes. This release and uptake of an H^+-ion was monitored by attaching a pH-meter to the membrane solution. Finally, they included data on the quantum yields of the photoreaction, that is, on how many photons were needed to catalyze the light reaction. The problem of quantum yield was a long-disputed issue in photosynthesis research, and in fact Hess and Oesterhelt used instruments from the recently dissolved Berlin institute of Otto Warburg to address the problem. Not only was Warburg's approach and position on quantum yield highly controversial, but his prewar apparatus versus minicomputer and double-beam photometer again exemplify a curious synchronicity of methods in this project that belong to very different ages of science.[200]

On the basis of dynamic, i.e., time-dependent spectroscopy, Oesterhelt and Hess argued that the photochemical changes were accompanied by protein "conformational changes," that is, rearrangements of the molecule's shape. Even if the overall approach of their work was still very much characterized by concepts from enzymology and physical chemistry, the question on the authors' minds was how to obtain insight into the "molecular mechanism" of the protein, as they put it.[201] That is, in this collaboration, we observe a gradual transformation from the enzymological approach Oesterhelt had practiced in Lynen's lab toward one of molecular biophysics that understands physiological processes as mechanical changes of proteins.

Meanwhile, Walther Stoeckenius had set up a separate group at NASA's Ames research facility at Moffett Federal Airfield. Among other things, Ames, located near Mountain View at the southern tip of San Francisco Bay, had been home to NASA's Exobiology Division since the early 1960s, and thus housed a number of scientists studying uncommon organisms such as *Halobacterium* by biophysical means.[202]

Stoeckenius used Ames to embark on similar experiments scrutinizing the purple membrane's photoreaction. The spectroscopic brand of biophysics needed for this task was completely different from his electron microscopic past, but it was the methodology that would become dominant in the publications of his group in the coming years. As he remembered, the Astrophysics Department at Moffett Field provided a laser, and Roberto Bogomolni, a young assistant professor whom Stoeckenius had hired from the group of photosynthesis researcher Melvin Calvin at Berkeley, set up a so-called flash spectrometer to study photoeffects with high time resolution.[203]

For this purpose, the purple fraction was cooled to −196°C, or liquid nitrogen temperature. The reasoning behind this was that the photoreactions' transient intermediate states, which vanished too fast to be observed at ambient temperature, could be "trapped" in the cold after a laser flash. When the scientists then increased temperature, the next step of reactions could be followed through changes in absorption of light. That is, manipulations of the purple substance's physical environment and experimental intervention with controlled light pulses rendered its molecular dynamics observable. Even if much more sophisticated, this strategy was in principle similar to the salt-ether-system that had rendered the color changes visible in vitro. The result was a model of BR's "photocycle," a scheme of cyclic light reactions, which explained the physicochemical changes the material underwent upon illumination.[204]

The transition from a concept of photoreactions inspired by enzymology as found in Oesterhelt and Hess' paper toward a graphic scheme of a photocycle was the next step toward a molecular-mechanical rendering of molecular function. The fact that the sequence of photoreactions was circular, leading back to the molecule's initial state, and that these reactions could be manipulated, interrupted, and re-started, may have contributed to the mechanical appeal BR gained at this time. This change in conceptualizing protein function is also indicated by the increasing frequency of the term "mechanism" in publications on BR in these years.[205]

Thus, BR research became connected on an instrumental and conceptual level with established fields of biophysics such as photosynthesis research, where spectroscopy and photocycles were the order of the day.[206] Spectroscopists' enthusiasm, which turned the curious find into a promising model for mechanistic protein studies, was related in no small degree to the fact that the protein was present as a pure and stable chemical substance—there was enough of it to fit into a cuvette for laser measurements; you could take the purple membrane on a train from Munich to Dortmund, or from San Francisco to Ames, handing it over to biophysicists not skilled in work with delicate enzymes. Here was a "part" of life's molecular inventory that was not only simple, but also robust enough to be studied.[207]

However, the elephant in the room of the purple membrane project must still have been biological function. In contrast to the suspected analogies with vision, and to all the data on the protein's photoreactions in the test tube, evidence of what BR or the purple membrane might actually do in a living cell was lacking. In other words, the biological raison d'être of these structures and molecules remained a matter of speculation. The reason was that studying living cells (the most obvious thing to do for most cell or microbiologists) had not been a priority for Blaurock, the structural biologist, nor for Oesterhelt, the biochemist, nor for Hess, the spectroscopist, or Stoeckenius, the electron microscopist. Everything until now had been done on preparations and molecules, with the living cell remaining a blind spot of the project.[208]

Cells in action—toward bioenergetics

In the summer of 1972, the entries in Oesterhelt's Munich notebook—still an example of thorough bookkeeping—became less frequent. The

FIGURE 9 The new molecular life sciences in West Germany. View of the Max Planck Institute of Biochemistry's building at Martinsried, south of Munich, undated; probably 1970s. The building's structure consisted of wings gathering loosely around centers, with each wing housing the institute's departments—this architecture was supposed to enhance interdisciplinary cooperation between these latter. In the back, supply buildings; in the front, next to the pond, library and guest house. Martinsried and the MPI became a nucleus for the formation of an entire science campus, attracting university departments as well as a biotechnology park and businesses since the 1980s (see ch. 4, Conclusion). Reproduced with permission of the Archives of the Max Planck Society, Berlin-Dahlem.

spectroscopic results from the collaboration with Hess were filed separately; moreover, the city may have offered ample distraction in the form of the Olympic summer games. But other reasons existed as to why less laboratory work was carried out: In these months, Feodor Lynen's Max-Planck-Institut für Zellchemie moved from the building shared with the university to a brand new research campus at Martinsried, back then merely a farmer's village south of Munich. In a major reorganization effort, the Max Planck Society (MPG) had decided to centralize Lynen's with two other Munich institutes, the Institute of Biochemistry, headed by Adolf Butenandt, and the Institute of Protein and Leather Research (fig. 9).

Instruments and laboratory equipment owned by the MPI were transferred to Martinsried, and the university department was left with few resources, as Lynen himself conceded. In his grant application to the German Research Foundation (DFG), Oesterhelt had already anticipated this situation a year previously, asking for support to purchase basic instruments.[209]

Still, his experimental options must have become very limited. With the benefit of hindsight, he remembered that the lack of sophisticated instruments (such as Lynen's automated amino acids analyzer) also had a positive effect—namely, the set-up of an extremely simple assay accomplishing what no one seemed to have done systematically so far: To study the physiological behavior of living *Halobacterium* cells in analogy to what was observed in the test tube.[210]

All that was needed for this assay was some standard glassware, a thermostat, a pH-meter, a powerful light source such as a projector lamp, and a culture of living cells. However, the stimulus to conduct these experiments (which could have easily been done two years previously at San Francisco) came from outside of Oesterhelt's ambit. He recalled presenting his data from spectroscopy and the proton reaction at a seminar at the Biozentrum in Basel, Switzerland. At this new, integrative center for the life sciences, membranes were among the topics *du jour*, yet again from a different angle.[211] In the discussion of what the protonation effect was about, a membrane biophysicist from Basel brought up the idea of trying to check for the proton reaction in the living cells with a pH-meter, that is, to see whether the same effect occurred that Oesterhelt and Hess had observed for a preparation. Upon Oesterhelt's reply that he could not do so since he possessed only a simple pH-meter, he insisted that this should be possible, quoting experiments on chloroplasts from photosynthesis research where such light-dependent changes had been recorded.[212]

The crux of this idea was to try to measure the same effect that had been recorded for a biochemical membrane for a physiological membrane, i.e., a functional part of a cell. Why did none of the protagonists from this story see before what was obvious to somebody involved in membrane research on photosynthesis? To understand this situation, we need to take into account the different groups researching "membrane-land" again—up to the early 1970s, membranes were something completely different for a biochemist than for a physiologist or a biophysicist and here we see one point at which these detached groups and their objects became connected.

The larger context of what the membrane researchers at Basel were doing was bioenergetics. What were bioenergetics' membranes around 1970? The central problem for those interested in how cells generated biological energy at their mitochondria or from photosynthesis at their chloroplasts was the membrane's contested role as a boundary to organize and catalyze biochemical reactions in space (see chapter 1). And right at the moment we are looking at here, the so-called chemiosmotic theory of British bioenergeticist Peter Mitchell was beginning to be ac-

cepted by more and more researchers—one advocate of Mitchell, Gottfried Schatz, was working at Basel, as the person stimulating Oesterhelt.[213] In order to experimentally scrutinize Mitchell's chemiosmotic
theory, which proposed that proton gradients across membranes were
a general driver of cellular energy generation, the alleged gradients were
measured using intact cells or cell-like organelles such as mitochondria
or chloroplasts. Mitchell himself had been using pH-meters to this purpose since the 1960s, determining the effect of oxygen on the rate of
respiration and hence energy production.[214] These oxygen pulse experiments, which Mitchell and his assistant Jennifer Moyle carried out on
isolated mitochondria, i.e., cell-like organelles, showed that the pH in
the test tubes decreased as an effect of a proton gradient building up
across mitochondrial membranes upon respiration, and hence increased
energy generation. As chemiosmosis became more and more accepted,
these experiments were considered as iconic for the new model to understand bioenergetics: A few years later, Gottfried Schatz would demonstrate similar pulse experiments in a video teaching session of the British
Open University, with the rapid movement of a pen recorder upon addition of oxygen to mitochondria supplying telegenic evidence for Mitchell's theory.[215]

As central as the issue of a proton gradient across the membrane as
the centerpiece of cellular energy generation may have been for bioenergeticists then, and as obvious as it is to biologists today, neither Oesterhelt nor Stoeckenius had come across it at the time. Underlining the
separation between the different membrane communities, Oesterhelt repeatedly asserted that he had not previously heard of Mitchell or chemiosmosis. In contrast to bioenergetics researchers, or physiologists, cells
had represented for him a source for biochemical materials or microscopic samples, but he had not researched them as intact, living organisms. Strikingly, his grant application from 1971 did not propose a single
cell physiological experiment.[216] Hartmut Michel, a later PhD student
of Oesterhelt and Nobel laureate of 1988, who had done internships
in Munich biochemical laboratories at the time, also remembered the
separation between enzymological or chemical biochemistry à la Lynen
and a more physiological biochemistry of the Medical School, where
chemiosmosis was a matter of discussion.[217]

The crucial and utterly simple experiment to put *Halobacterium* cells
encompassing the purple membrane into a test tube and check whether
the pH of their medium decreased upon illumination must have been out
of the ordinary for Oesterhelt. Whereas he dated, numbered, and described his other experiments neatly, the scale paper slips documenting

the pH-pulse measurements on living cells were documented on a different type of paper, were less comprehensively described, and remained
frequently undated.[218] Filed separately, they may have been considered
as preliminary, and/or they may have been carried out beyond the usual
confines and rhythms of lab work.

From the cross-references between the labels of the cell cultures and
the dates of their production in his main notebook at the time, the approximate onset of these experiments can be dated to summer 1972.[219]
Oesterhelt remembered these first experiments vividly, and framed them
in the narrative of an arcane discovery: In a dark room of the basement,
he would have filled cell suspensions in a glass cuvette, attached the electrode of a pH-meter to the solution, and set up a projector lamp.

The pH meter's sensitivity was adjusted to the maximum, and when
he switched on the light, the recorder's pen jumped to the edge of the paper slip with an unexpected intensity: The biological activity of the purple membrane became tangible in the simplest set-up, and the graphic
method that was a tool of the trade not only in bioenergetics, but in
physiology generally rendered it perceivable in an impressive way.[220] At
the basis of this effect was again the active membrane: Now it was not a
substance materialized in the test tube, but a part of living cells, present
as a violet-colored culture in a glass cuvette, that *reacted* to illumination,
generating a spike on paper. This immediately suggested that the membrane's light reactions were part of the organism's energy metabolism,
and this turned the purple membrane into an object of photosynthesis
research and therefore bioenergetics. So, the idea from Basel instanta-
neously put the effects and reactions of the purple membrane on a whole
new stage (fig. 10).

The manifold recollections I have read and heard about these pH-
pulse experiments, which were adopted by many in the early 1970s BR
community, and then turned into a demonstrative experiment for teaching, have prompted me to repeat it. Using the gear lying around in a contemporary biology laboratory, and a leftover culture of *Halobacterium*
cells, I had no problem to reproduce curves similar to Oesterhelt's with
some basic lab knowledge. Bearing in mind the complexity and fickle
set-up of most biochemical experimentation, it was remarkable how easily this physiological process could be recorded, and how robust the assay was. As cells actually seemed to do something upon illumination,
creating repeated motion of a pen, the experiment involves a high level
of sensory experience, with "bioenergetics" getting a quite literal meaning beyond its complex theoretical background.

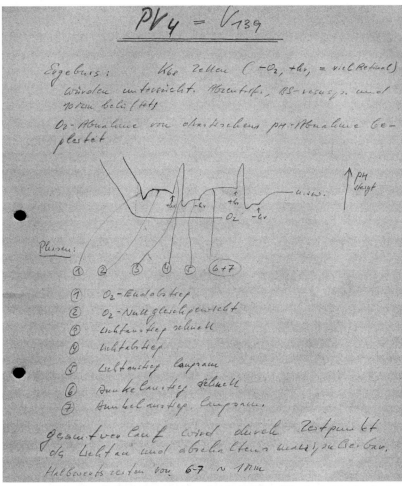

FIGURE 10 Energetics cells. Handwritten sheet from Oesterhelt's folder of his pH-pulse experiments, displaying the membrane activity of living cells upon illumination, 1972–1973. Undated graph of a pH profile as generalized from several experiments with a pen recorder attached to a customary electrical pH meter. Numbers indicate different phases of the experiment. After the light is switched on (indicated by "+h*v", phase 3), a transient increase of pH in the cuvette is followed by decrease (phase 4), which was later taken as the main effect of bacteriorhodopsin: Pushing protons out of the cell, acidifying the medium. After switching off the light, pH returns to baseline (phases 5–7). This diagram also illustrates the adoption of physiological experimentation and the graphic method (as used in bioenergetics at the time) into the new molecular membrane project. From Folder "Protonenversuche/Bleichungen" (see note 218). Reproduced with permission of D. Oesterhelt, Martinsried.

There is another reason as to why these pH pulse experiments were startling for a biochemist around 1970, as illustrated by the reported disbelief of Lynen when Oesterhelt told him how he interpreted his experiments—that bacteriorhodopsin's function may be to transfer or pump protons across the cellular membrane.[221] The observed pH changes must have appeared out of the ordinary since they were repeatable. Usually, pH changes in solutions, for example, after two substances have been mixed, went one way: Once the chemical reaction reached equilibrium, the pH would remain constant. Here, the observed reaction was auto-reversible and could be reproduced after the pH had returned to the initial level during a rest phase of the cells in the dark. Shine a light, pH decreases; switch it off, pH returns to normal; shine it again, pH decreases again. Similar to the reversible color change in the test tube, the reaction appeared akin to *mechanical processes* rather than to ordinary *chemical reactions*. After some repetitions, however, the cells in the sample seemed to lose their power, as shown by a faster return of the pH to the baseline—cellular activity "weakened," as known from biological processes such as muscle action.

At a meeting of the German Gesellschaft für biologische Chemie (Society for Biological Chemistry) in October 1972, Oesterhelt presented his spectroscopical results from the salt-ether-assay. When pondering the function of the protonation effect in vivo, he ventured an interpretation of his first pH-pulse recording from July. The purple membrane would thus serve the microbes to build up a proton gradient across the membrane upon illumination, similar to what was discussed in chloroplasts—the possible coupling of this process to cellular energy production from light, and thereby the significance of this new object to bioenergetics, could easily be grasped.[222] Since the fall of 1972, the pH pulse experiments must have spurred Oesterhelt's interest, as he continued experimenting with varying conditions such as measurement times, temperature, and light intensity. The spikes of cellular activity he recorded became more differentiated, and in fact, the main direction of the process was found to be opposed to what he had first presented, i.e., a net proton *export* from the cell occurred in *Halobacterium* rather than an *import* as in chloroplasts.

At the time, Oesterhelt and Stoeckenius frequently exchanged their results via mail, and Stoeckenius went to Munich presumably in September. A letter from the following month makes first mention of pH-experiments carried out in California as well, and here, the first mention of the purple membrane as a "light-driven pump" was made.[223] By that time, Stoeckenius had reached out to the Berkeley lab of photosynthesis

researcher Melvin Calvin. Calvin's former PhD student, Roberto Bogomolni, recalls having performed experiments on a possible connection
between BR and photosynthesis at Berkeley and was soon after hired by
Stoeckenius. The development of the pH-pulse experiments toward their
final form published in 1973, and their interpretation occurred in close
but complex interactions between these individuals, mingling cooperation with competition, as documented by retrospective publications and
correspondence. [224]

Under the title "Functions of a New Photoreceptor Membrane," the
energetic spikes of the purple membrane were published in the October 1973 issue of the *Proceedings of the National Academy of Sciences
of the USA*. Here was an answer to the question that had been the elephant in the room, biological function, but it was different than expected: The membrane's "bleaching" or photocycle was not associated
with light perception, but with proton transport across the cell membrane. This process was seen as a means for the cells to generate energy,
and thus the purple-red protein's function appeared similar not to the
eyes of animals, but to the green membrane-bound protein, chlorophyll,
and thereby plant photosynthesis. On the basis of pH-pulse traces similar to those found in later phases of Oesterhelt's records, the paper argued for the "direct pumping action of bacteriorhodopsin."[225]

Subsequent to these cell physiological experiments, the purple membrane would represent what the 1973 paper described with an energetic
metaphor that should become momentous, omnipresent, and unques-
tioned—a "pump."[226]

The relevance of this paper may result from the fact that this simple,
tractable protein was connected to two central topics related to membranes at the time. First, this was physiology of nerve and muscle, where
the existence of elusive "pumps" to build up electrochemical gradients
had been discussed since the 1940s (see chapter 1). The second connection, to bioenergetics and chemiosmosis, was made explicit by referenc-
ing Mitchell's 1967 paper on oxygen-dependent pH-shifts in mitochondria. BR as a light-driven ion pump not only seemed to bind together on
a molecular level conceptually different biological processes such as light
perception (through the chemical similarity with rhodopsins), photosynthesis, and generation of electrical potentials, but—taking into account
its easy accessibility and handling—it lent itself to address many of the
pressing issues related to membranes. Max Delbrück, who seems to have
been frequently briefed about the purple membrane, admitted to Oesterhelt in a letter from September 1973:

Dear Dr Oesterhelt: At long last I got around to reading your magnificent four preprints with some care, and to discuss them with Lily Jan. You have certainly accomplished in a short time an impressive amount of very high quality work on this fascinating system. It puts me to shame when I think how slowly we have been progressing over the years with our Phycomyces problems.[227]

That said, he enclosed a meticulous three-page critique of one of Oesterhelt's manuscripts. This most tangible example of an active preparation, a macromolecule that could be isolated, and the effect of which could be studied in living cells, promised to indeed be nature's "clue" to get a molecular grip on membranes.

Plugged into the circuit—a "molecular electric generator,"
Moscow 1974

The pump was not the only technological concept applied to BR in the early 1970s. Bioenergeticists from the Soviet Union, who had rapidly taken up the new object after the first publications, modeled the light-dependent effects of the protein as an electrical generator on the molecular level.[228] To accomplish this, the team from Moscow State University set up an experimental assay that accomplished another conjuncture of electrophysiology and biochemistry by integrating the new material substance's effects in an artificial model membrane (created by a septum of a nitrocellulose film), to which BR-containing liposomes, microscopic spherical lipid vesicles containing patches of the purple membrane, were adsorbed (more on liposomes and similar cell models in chapter 3). When these thin films were illuminated, the electrical current produced by the protein was detected and recorded by the electrodes (the movement of a positively charged proton also generates an electrical effect). As the curves displayed BR's action in millivolts, it was only logical that the Soviet group represented their assay and its components in the form of a circuit diagram as known from electronic technology or cybernetics— BR with the sign for an electrical generator, the membrane as a capacitance and a resistor, coupled via electrodes to the voltmeter.

Both the assay and the modeling of membranes as electronic technology harked back to Hodgkin and Huxley's model of the nerves' action potential, and further into interwar biophysics (chapter 1). New, however, was that these experiments were able to pinpoint the electrical phenomena down to the molecular level. This, in turn, demonstrates how bioenergeticists and physiologists took up the new object in order to finally render membrane phenomena such as energy generation or neuronal signal transmission molecular. Thus, this assay took a similar direction to an epoch-making technique of neurophysiology that dates also to the mid-1970s—"patch-clamping," which allowed physiologists to measure the electrical effects of single protein channels in tiny membrane fragments of nerve, developed by the German biophysicists and physiologists Erwin Neher and Bert Sakmann.[229]

The 1970s experiments shaping BR as an electrical generator should be conceived of not only as a way to molecularize neurophysiology, but also as part of a long strand of modeling biological activity in terms of electrical technology, reaching much further than analogies on paper. In the Soviet work, as much as in a number of follow-ups from the US or Germany, the purple membrane literally turned into a functional biological molecule plugged into a circuit, and thus an object in between the molecular life sciences and physico-chemical technologies.[230] In the wet environments of cells and test tubes as much as in the dry environments of electronics, the molecule was able to achieve effects that suggested it may eventually be able to do technological jobs—generating energy, or switching currents. So, these electrophysiological studies paved the way for a development that was to kick in a decade later—the schemes and ideas to create a new, life-inspired molecular technology that some thought would revolutionize microelectronics (chapter 4).

What is more, the paper from Moscow also hints at a possibly different geography of membrane research as compared to the largely "Western" story of molecular genetics: Whereas the Soviet Union had clearly experienced a lag in terms of genetics due to the effects of the Lysenko affair, other molecular biologies had not suffered. Among these were intermediary metabolism biochemistry, muscle research, as well as bioenergetics.[231] As a new "molecular biology without genetics," membranes and bioenergetics thus met with existing biochemical and biophysical expertise. Moreover, the discovery of BR in the early 1970s fell into a period when the molecular life sciences received a boost in the Soviet Union, in which the academician and later member of the communist party's central committee, Yuri Ovchinnikov, played a significant role.[232] Ovchinnikov's own field was organic chemistry and analytic biochemistry, such as the purification of biologically active molecules (antibiotics, peptides, or toxins), many of which acted upon biological membranes. BR seems to have made such an impression on him that he set his own group to work on it. In combination with Munich, and other places of BR research that would soon be added to the map, e.g., in Israel or Hungary, the geography of membranes in the 1970s thus suggests both a different genealogy and a global extension of the molecular life sciences beyond genetics.

To return to the BR story until around 1975, the increasing use of technomorphic terms such as "pump," "electrical generator," or "photocoupler" in connection with the new protein indicates a development of the field away from its formation at the crossroads of enzymology and membrane structure toward a molecular-mechanical perspective, which hinged on spelling out protein dynamics by spectroscopy, electrophysiology, etc. A similar transformation affected membranes and bioenergetics throughout, with the phenomenological perspective of physiology (centering on bulk movements of molecules, as in Hodgkin and Huxley's model of the action potential) displaced by microphysical models of "what happened" within one pump or channel molecule during function.[233]

The pump takes shape, Cambridge 1973–75

Since the 1980s, the molecular-mechanical picture of membrane processes has become dominant, with the pump BR or the ATP-synthase as their poster children. In 1974, however, these issues were still very much open to debate—not least since *no* structural model of a membrane protein comparable to DNA, hemoglobin, or other enzymes had been established. In spite of increasing attention, BR was still very much an "object in flux," as one could say, to modify an expression of Yehuda Elkana.[234] It was unclear, for example, if chemical effects and reactions related to the protein molecule, or the entire membrane. Moreover, it was an open issue if the functional object in the cell correlated to the protein material in the test tube, or how BR was imagined to "sit" in the membrane film. These and more questions were basically unresolved for all receptors, transporters, and channels, from nerves to bacteria. However, many of them were settled quite abruptly around 1975 on the basis of a visual model of BR, and the story of how this was established with a novel method of structural biology will be told in this section.

The molecular basis of membrane proteins and their dynamics were a pressing issue in the early 1970s. Recall the statement of molecular biologist Arthur Pardee that, at this level, processes central to metabolism, immunology, or neurophysiology remained "completely mysterious," or recall Jardetzky's speculative sketch of how such pumps were supposed to work. The absence of a visualized molecular model must have appeared striking against the successes of structural biology and its richly illustrated textbooks, such as Dickerson and Geis' *The Structure and Action of Proteins* (see chapter 1).

One young protein crystallographer who took up the challenge of membranes was Richard Henderson. With a degree in physics, Henderson had done his PhD work in David Blow's group at Cambridge's LMB, a hub of X-ray structural biology, where he had contributed to the elucidation of the molecular mechanism of the digestive enzyme chymotrypsin in a project that brought together structural biology, enzymology, and protein sequencing.[235] During his postdoc at Yale, Henderson's aim had been to select another enzyme for X-ray work, possibly from intermediary metabolism. However, as he remembered, his mentor suggested a different strategy: As a young scientist, he would be better to choose something that would make an impact in 20 years' time.[236] Thinking back to Cambridge biophysics, Henderson came up with the idea of trying to solve the structure of the membrane-bound sodium channels of nerves. The molecular basis of the well-researched action potentials had remained an unresolved issue. *Grosso modo*, the "nature of the molecular events underlying changes in permeability" that had eluded Hodgkin and Huxley in the 1950s, were still beyond reach.[237] There was some recent development on the issue, however, as Jean-Pierre Changeux or Ricardo Miledi from the then coalescing field of "neurobiology" were gradually getting a biochemical hold on receptors from the synaptic membranes of nerves. They purified proteins from their membranes as material correlates of receptor action, using for this purpose attached toxins or radioactive labels.[238]

In 1970, Henderson embarked on a similar approach to the alleged channels in nerves, synthesizing small molecules inhibiting their action, and radioactive labels. These were to be used as "tags" in order to isolate the proteins from the membranes by methods such as chromatography, or to stabilize them for crystallization.[239] Remember that biochemical protein isolation was well-established for water-soluble enzymes, but remained troublesome for membrane proteins, as these tended to "stick" within the colloidal lipid-protein mix of the membrane. However, large amounts of pure protein were a precondition for crystallography. As Henderson was toiling with such sophisticated labeling strategies in order to isolate Hodgkin and Huxley's channels from nerve membrane preparations, it is no wonder that a plenary given by Walther Stoeckenius at a San Francisco biochemical meeting in June 1971 caught his attention: Here was a membrane protein of which one could get large amounts in a pure and crystalline state, and which seemed to be related to the visual pigment. Also to Henderson, the early reports on BR must have sounded as if nature had offered a material clue on the membrane problem.

After a frustrating year and a half spent with preparatory work on sodium channels, he decided to switch to this easy example of a membrane protein. He remembered convincing Don Engelman, a membranologist at Yale who had published with Stoeckenius, to phone the latter in San Francisco, to inform him of Henderson's plans and ask for a cell culture. Stoeckenius did not want to collaborate, but agreed to send cells for Henderson to start on his own. The correspondence between Stoeckenius and Henderson, the latter a junior of Oesterhelt's generation, documents the difficulties of setting up the new project's infrastructure. As a structural biologist, Henderson had no prior knowledge of cultivating microbes, similar to Stoeckenius almost a decade earlier. Thus, he tentatively asked Stoeckenius for some more details on procedures and the state of his work, which he received. Letters between the two researchers were frequent and collegial in tone from the beginning, and soon, Henderson was returning his own observations. Only a year after he had started—Henderson was back in Cambridge by then, and had taken the new topic home (also similar to Oesterhelt)—the progress of his work may have convinced Stoeckenius to offer this potentially successful rival a position as a structural biologist at UCSF. Now, it was Henderson who hedged his bets, however, stating that he had just settled in Cambridge and preferred to carry out his plans there for now. He enclosed new X-ray pictures he had taken, which Stoeckenius found "impressive." In 1974, he finally declined the offer to form a structural biology group in San Francisco, for the stated reason that a critical mass of instrumentation and staff to continue working as he had been in Cambridge were not present.[240] The correspondence illustrates the shifting relations as well as the exchange economy of these two scientists who were working on the same problem, with their reciprocal needs, the changing differences in their social status, and their interest in maintaining independence at important points.

As Stoeckenius' former postdoc Allen Blaurock was also continuing to work on purple membrane structure at King's College, the situation became even more complicated. Henderson's work proceeded well, and with a Cambridge researcher having taken up a structural topic that someone was pursuing in Maurice Wilkins' department at King's, the possible outcome for Henderson risked sounding like an echo of the DNA story two decades earlier, when Jim Watson and Francis Crick had used Rosalind Franklin's experimental data without permission or giving due credit. Thus, having submitted a manuscript on X-ray diffraction data on the purple membrane to the *Journal of Molecular Biology* in 1974, Henderson recalled driving down to London to discuss his work with Blaurock and passing him the manuscript—the papers by the two authors then appeared back-to-back in 1975. The somewhat ironic point about this resolution of what could have become a conflict in light of past events was that neither of these two X-ray structural papers was later seen as *the* breakthrough in the problem of membrane protein structure. This only came about when Henderson tried out a new method.[241]

Material bricolage

At the LMB's Annual Laboratory Symposium in October 1973, an internal gathering to share each other's work, Henderson remembered hearing an electron microscopist of Hugh Huxley's muscle structure group, Nigel Unwin, talk about electron microscopy of the tobacco mosaic virus. Due to its repetitive assembly from helically arranged protein subunits, the virus had been used as a model object by structural biologists such as Rosalind Franklin. The virus project had started even before the double helix and was continued after Franklin's death by Aaron Klug and others.[242] Henderson noted Unwin for a curious methodical invention: It may sound anecdotal, even outright bizarre, but Unwin had tinkered with a gold-coated spider thread inserted into the instrument's electron path. The filament served as a phase plate, that is, a radiation-permeable material modifying the beam's phase. The phase contrast approach was known from 1930s' light microscopy, where it had allowed an important enhancement of resolution.[243]

To understand where Unwin was aiming with his makeshift phase-contrast electron microscope, it is necessary to briefly recall the power and the problems of the approach. Since the 1950s, electron micrographs of biological material were usually produced from samples stained with heavy metal salts, which, however, brought with them the risk of producing artifacts. A phase-contrast technique promised to detect the low contrast produced by the electron deflection of the biological material itself, which would mean that one could omit the critical staining process.[244]

Unwin's gilded cobweb, surely somewhat fickle, never came into play for the purple membrane, as an easier trick to accomplish the same effect existed (it may have been merely a catchy episode of the talk). Yet, in a sense, the cobweb strategy is illustrative of the project Henderson and Unwin had started: Leaving behind the established methods

FIGURE 11 Membrane electron microscopists. Richard Henderson (right) and Nigel Unwin on the stairs of the old building of Cambridge's Laboratory of Molecular Biology, 1980s. Unwin worked on neuroreceptors after the joint development of crystallographic EM on BR in the early 1970s. Reproduced by courtesy of the Medical Research Council–Laboratory of Molecular Biology (MRC-LMB), Cambridge, UK.

of X-ray crystallography and EM, the two developed a merger of both, which involved material bricolage, or the makeshift assembly of resources that were around in a novel way (fig. 11).

Nigel Unwin apparently had a predilection for this work style. After obtaining a PhD from Cambridge's Department of Metallurgy, where he had probed the composition of metal surfaces by EM, he was hired by Huxley to the LMB in 1971. Here, Unwin tried out novel ways to conduct EM on model specimens and tried out methods for sample preservation. Thus, such an ample and apparently well-structured substance as the purple membrane must have fit with Unwin's methodical interests— the material promised to spend time trying out methods rather than obtaining it.[245] However, even a material as ample and rigid as the purple membrane deteriorated under the harsh vacuum conditions required for EM. When the team observed that contaminated samples actually produced better images than pure samples, they took up Unwin's experiences with different additions to the specimen. In a sense mimicking contamination, they added conventional sugar to the sample and ended up with something that one may describe as a "sugar icing" of the membrane.

Instead of using synthetics or highly reactive heavy metal salts, sample preservation was achieved by tinkering with simple substances found in any lab, thereby emulating processes known from a kitchen or candy shops.[246]

The first joint trials of phase-contrast EM with an unstained purple membrane specimen worked out well, as Henderson vividly remembered:

> And then looking down the binoculars with your eyes, no film, nothing like that, you could see the spots really brightly on the screen shining at you. So you can see the diffraction spots. At that point we knew—it is like three weeks after we had started fiddling around—we knew it would work out [. . .].[247]

As in many of the other early experiments, the purple membrane became an attractive object of study for Henderson and Unwin because it immediately produced tangible results—you could see with the bare eye what was impossible to perceive in other cases. The material *lent itself* to being worked with in many fields of the life sciences, and it advanced to a model allowing novel methods to be pioneered and applied in a short time. The effect of this was that it connected researchers from different fields, which used different techniques—and in turn many of them would soon work on the same object.

Data instead of images—a new electron microscope

Hardly anything can be discerned in what Henderson and Unwin qualify as "typical" short exposure micrographs of the purple membrane.[248] The delicate balance between recording an image and destroying the sample by the very electrons required for imaging resulted in more or less uniform, "underexposed" gray plates. These were interspersed with black shades resulting from gold particles, so-called gold islands, which were added to the carbon grid sample holder, providing something like "landmarks" to spot the position of the sample on an otherwise fuzzy screen.

How did Henderson and Unwin proceed from these images that showed almost nothing to *enhancing* resolution of the electron microscope, thereby pioneering a novel use of the instrument toward visualizing macromolecules that has nowadays become another pillar of structural biology?

The explanation is that the duo treated electron micrographs not as *images* to be inspected by the eye, as in light microscopy, but as *data sets* displaying reflection patterns and intensities similar to those from X-ray crystallography, that were analyzed mathematically in order to reconstruct images. Aaron Klug, student of John D. Bernal's famed crystallography department at Birkbeck College, London, and collaborator of Rosalind Franklin on the structure of tobacco mosaic virus, had developed principles of this method, and as group leader at the LMB, he was in the ambit of Henderson and Unwin. Klug explained the procedure of what came to be known as crystallographic EM as follows: In a first step, a Fourier transform of each single micrograph was produced; that is, the image was turned into a data set by quantifying the optical densities of the spots seen on the micrograph (resulting from diffraction of the sample) and processing these mathematically with a computer. When this had been done for a great number of micrographs, the back transformation of the obtained Fourier-coefficients allowed an image of the object under scrutiny to be reconstructed, in which the noise, or the blurriness, of the single shots was averaged out. One big advantage of this method of structure determination was that no large three-dimensional crystals were required; two-dimensional sheets such as the purple membrane would suffice.[249]

The need to turn micrographs into data sets meant in practice that Henderson and Unwin had to quantify the gray scales of numerous large image plates, a task similar to the film scanning routines in the LMB's X-ray crystallographic projects. This laborious and error-prone work had previously been carried out by unskilled, often female laborers—the so-called computer girls—but by 1972, the institute possessed a semi-automated film scanning device, or microdensitometer.[250] However, the machine was found to be unsuitable for electron micrographs and the two had to turn elsewhere. Henderson filled the trunk of his car with film cassettes and drove around London to Herstmonceux, Sussex. Here, at the Royal Greenwich Observatory, an automated scanning densitometer with high resolution existed for the evaluation of astronomical photographs, which he was permitted to use at weekends. Recorded on magnetic tapes, the image data were then processed on an IBM computer in Cambridge's Mathematical Laboratory, one of the facilities the LMB made use of before their own computing power became sufficient for such calculations in the 1980s.[251]

Henderson and Unwin's approach was "radical" (thus Aaron Klug) not only because unstained specimens were used, but more generally because it tied together an unconventional object for EM, experimental bricolage, and a new way of dealing with images as data sets, which involved different resources and actors within and beyond the LMB.[252] It is hard to imagine how such a methodical tour de force could have worked outside of the few centers of structural biology and EM that then existed in the world, which brought together the necessary devices and skills as well as theoretical expertise. However, the approach was also radical because it foreshadowed a novel use of an instrument and a novel concept of an image: Electron micrographs treated as arrays of data, read by machines rather than inspected by the eye, were a profoundly different way to employ an instrument that had become customary in the life sciences since the 1950s. In a more extended historical perspective, this was one step away from an EM as a visualizing method in analogy to light microscopy or photography to a method that produced reconstructed images. This repurposed, different electron microscope became the topic of Henderson's work for the coming decades, and it in fact changed the instrument's overall use dramatically, which stands in our times next to X-ray crystallography as another method of choice for protein structure determination.

Contouring the pump

Unwin and Henderson described their new approach in a 1975 issue of the *Journal of Molecular Biology*, the purple membrane being one among other proteins to demonstrate the technological innovation, rather than the focus of the argument. However, the images from this work depicted the heretofore unseen dimension of membranes in between molecular and cellular structures (fig. 12).[253]

To pick up Wolfgang Ostwald's slogan, the new electron microscope allowed one to get a glimpse of the "world of neglected dimensions," right at the time when a synthesis of data accumulated in different fields had finally led to a novel structural model of membranes as a fluid mosaic, that is, as dynamic, self-organizing double lipid films containing domains of proteins spanning them from face to face or sitting in them.[254] The delimitations of molecular objects, one of which was the protein BR, the others the far smaller lipids, were traced and modeled by contour lines indicating the density of electron scattering matter. Quite literally, the analogy with geographical mapping suggested a "molecular landscape"—the membrane as an ordered space in which metabolic processes were to take place. At a resolution of 7 Ångström (a unit equal to a tenth of a nanometer), which was still far away from what could be achieved by X-ray crystallography, the protein's boundaries and its substructures emerged as cylindrical features perpendicular to the membrane plane.

The ensuing *Nature* paper by Henderson and Unwin can truly be said to have changed the concept of membrane proteins. From two-dimensional

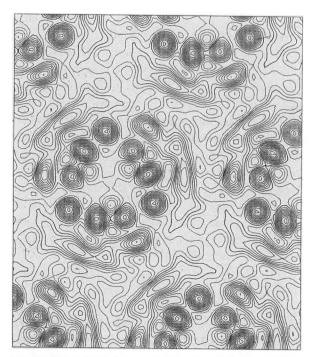

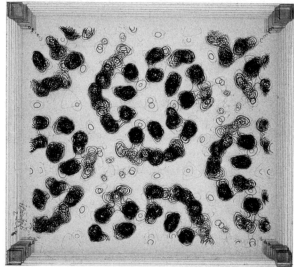

FIGURE 12 Molecular landscaping of the membrane. Top: Plot from Richard Henderson and Nigel Unwin's joint work in 1974. A three-color plotter from Cambridge was used to generate a top view of electron density (indicating the distribution of scattering matter) within a purple membrane patch. Such maps resemble geographical contour lines. Three round elements plus the adjacent curved shape were identified as the seven alpha-helices of one bacteriorhodopsin molecule, oriented perpendicular to the membrane plane. Three such molecules were oriented around one symmetry axis. To reconstruct a 3D model of the molecule, Henderson and Unwin transferred such cross-sectional plots from different levels of the membrane to perspex plates, as displayed in bottom figure. Stacks of these plates allowed them to reconstruct the spatial contours of the BR molecule as shown in figure 13. Reproduced by courtesy of the MRC-LMB, Cambridge, UK.

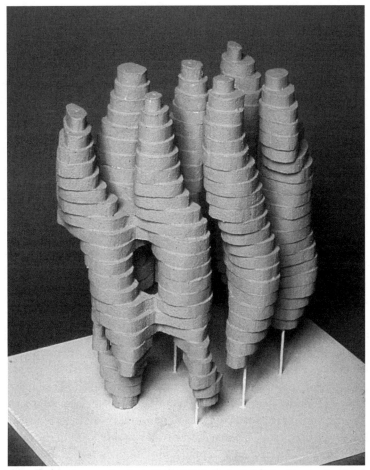

FIGURE 13 A molecular pump visualized. 3D-model displaying bacteriorhodopsin's seven transmembrane helices as shown in Henderson and Unwin's 1975 *Nature* paper. The membrane plane is perpendicular to the helices. This was the first structural model of a membrane protein, and it was established with a new method. However, the model's accuracy, or technically speaking the resolution of the electron micrographs used to establish it, was low in comparison with what X-ray crystallography accomplished at the time— protein substructures were not visualized. Reproduced by courtesy of the MRC-LMB, Cambridge, UK.

images, they had advanced toward a three-dimensional model of the purple membrane and its protein, which came even closer to what crystallography accomplished.[255] In practice, the information required to model the third dimension was gathered by tilting the membrane specimen with respect to the incident electron beam. The molecular structure of the membrane, as a crystalline arrangement of proteins, was then literally assembled by stacking transparent plates on top of each other.

These plates indicating electron density profiles represented different depths of the membrane film (fig. 13; for explanations see also fig. 12).

The overall appearance of the protein's structure, or so the authors continued, was characterized by "numerous rod-shaped features aligned perpendicular to the membrane."[256] Biochemically speaking, these features were known as "alpha-helices," a protein substructure that Linus Pauling had previously described. The *Nature* paper culminated in a photo of the protein model, assembled from balsawood slabs jigsawed according to the electron density lines by the LMB's workshop. Graded, slightly twisted columns represented bacteriorhodopsin's seven alpha-helices gathered around a central pore. The model may appear coarse-grained with respect to the drawings of molecules in Dickerson and Geis' book, but at the time, it was a big step: For the first time, a membrane protein had become tangible, modeled as a thing, which one could not only look at, but take into one's own hands and turn around. Here was a molecular pump—materialized and visualized.

Visualizing molecules and mechanisms

With respect to the membrane problem that troubled physiology, bioenergetics, photosynthesis, or pharmacology, the 1975 BR model provided evidence for answering at least two pressing issues. First, regarding membrane architecture: In contrast to what Davson and Danielli had postulated four decades previously, and in line with the new fluid mosaic model, the membrane formed a two-dimensional "landscape" in which proteins and lipids were arranged neatly side by side, with the hexagonal arrangement of BR molecules confirming the membrane's mosaic-like architecture.[257] The pump protein sat in the membrane and traversed it from one side to the other, rather than being attached to either face of it. Second, regarding the function of membrane proteins: BR was presented as a "simple example" of other pumps and channels, such as those discussed by Hodgkin and Huxley, the existence of which had so far been based largely on indirect evidence. The visualized pumping molecule suggested how other such proteins would have to be imagined and how they could function as mechanical devices on the molecular scale.

Looking at the image, it may not be immediately evident from this static model how the molecule would accomplish its biological function of pumping protons. Yet, the idea of a protein straddling the membrane from one face to the other rendered some explanations that abounded at the time less plausible than others: Whereas it seemed now difficult to imagine that the protein rotated within or diffused through the membrane, it was much more likely that transport was accomplished by re-arrangements or conformational changes of the molecule, that is, by mechanical motions of the protein, such as proposed in Oleg Jardetzky's schematic model of 1966.[258] Bending or twisting of its alpha helices, the seven columns traversing the membrane, made protein function conceivable as simple mechanical movements, as opening, closing, pushing, etc., akin to what we know from interactions with macroscopic things of our everyday world. Mechanical dynamics of molecules, discussed also in spectroscopy (see above) and conceived of as rearrangements of assumed rigid structural elements, would bring about physiological processes, such as moving a proton "freight" across a membrane. This molecular-mechanical thinking—presuming in a way a determinist picture of the molecular world—became pitted against a statistical conception of the problem inspired by thermodynamics and physical chemistry (as found in Peter Mitchell's conception of membrane dynamics and chemiosmosis; see chapter 3).

Henderson and Unwin's model formed a center of gravity for future research, and in consequence, the research questions of ensuing projects changed: The spatial organization and temporal dynamics of the macromolecule and its helical substructures within the membrane moved into the focus, with conformational changes becoming related to function. The ultimate goal became to spell out a "molecular mechanism" of this molecule.[259]

Taking into account the explanatory power and the promises of molecular models, as well as their inherent analogy between macroscopic and molecular processes, it was only a small step for scientists to conclude that membrane proteins actually *were* mechanical devices on the molecular scale. Thus, in terms of impact, the small 1975 balsawood construction ranges among other models of "living molecules" (the title of a long-running exposition at London's Science Museum) that changed the image of life at the molecular scale, such as Watson and Crick's DNA model, Rosalind Franklin and Aaron Klug's tobacco mosaic virus, or John Kendrew and Max Perutz' myoglobin and hemoglobin proteins. Photos and drawings of Henderson and Unwin's model were shown and reprinted multiple times in the scientific and nonscientific press, and one of the balsawood constructions is today kept at the Science Museum.[260]

Toward cryo-electron microscopy

The story of structural research on the purple membrane did not end in 1975. Whereas Henderson and Unwin's model had contoured the pump, and thereby helped to answer a number of open issues regarding membranes, the model's resolution was far from rendering protein parts such as amino acids, let alone single atoms, visible. In the coming twenty years, scores of scientists would engage in the quest for high-resolution structures of this and other membrane proteins.[261] Richard Henderson pursued the electron microscopic approach at the LMB, with BR becoming his model object for method development and enhancement of resolution. An important step in the long-term endeavor to establish EM as another method to determine protein structure (beyond X-ray crystallography or magnetic resonance spectroscopy) was the set-up of a consortium of research groups under the umbrella of the European Molecular Biology Organization (EMBO), comprising the LMB, the EM department at the MPG's Fritz-Haber-Institut at Berlin (FHI, headed, after the retirement of the developer of EM, Ernst Ruska, by Elmar Zeitler), as well as industrial partners such as Philips.[262] In the 1980s, this consortium used the expertise of the different groups in instruments development to devise crystallographic EM methods that worked at liquid helium temperatures in order to minimize thermal vibration and radiation damage. The purple membrane, as other supramolecular biological structures such as ribosomes, were studied by these early cryo-electron microscopes, huge and sophisticated prototypes housed in shock-absorbing towers next to the FHI.[263] The outcome of this research and development consortium was not only a refined EM structure of BR that stretched down to the level of atoms, published in 1990, but more importantly technical and conceptual contributions for a method that has become a viable alternative for molecular structure determination. Single-particle cryo-EM (which does not require any ordered arrangements of molecules such as crystals or membranes) is nowadays applied to an increasing number of proteins, and has, on the basis of improved optical technologies, led to a "resolution revolution" in structural biology.[264] In December 2017, as this chapter found its present form, Henderson's contribution to the development of cryo-EM has been rewarded by a share of the Nobel Prize in chemistry. Whereas his co-awardees, Jacques Dubochet and Joachim Frank, contributed rapid freezing methods and mathematical methods to treat the image data, respectively, the Nobel committee mentioned the BR work in Henderson's case. Maybe one could take this statement one step further and add that the material substance that had caught his attention in the early 1970s had an impact on redirecting him from crystallography to this new field—as he remembered in an interview, the BR's spots seemed to directly shine at him through a binocular in the first trials, and that it was this object that made the collaboration with Nigel Unwin fruitful.[265]

Henderson, to be sure, was not the only structural biologist who noted the material promise of the purple membrane and took a bet on it: Hartmut Michel, PhD student in the lab of Dieter Oesterhelt, stumbled upon crystal-like formations of BR in a preparation in 1977 — a trope repeatedly mentioned in crystallographers' stories, underlining the serendipity involved in obtaining protein crystals.[266] Thus, Michel, doing Mitchellian bioenergetics studies of BR before, re-directed toward structural biology, visited Henderson at Cambridge and attempted to resolve the purple pump's structure by X-ray crystallography. To make a long story short: It did not work, with BR remaining recalcitrant to X-ray studies until the late 1990s, but Michel, who became a group leader in Oesterhelt's department at the MPI of Biochemistry in the 1980s took up another microbial membrane protein that naturally formed crystals. From this photosynthetic reaction center, he and crystallographer Johann Deisenhofer resolved the first X-ray structure of a membrane protein, sharing a chemistry Nobel with Robert Huber in 1988.

While a more detailed history of structural biology after 1970 showing the contributions of and displacements between different methods such as X-ray, cryo-EM, or NMR remains to be written, it is obvious that the purple membrane as a material that lent itself to structural studies would play a significant part within it, and that research on it helped to shape the present situation, in which protein structure determination is increasingly becoming routine research. In the wake of the 2017 Nobel prize, it was suspected that resources may be redirected from the large synchrotrons producing X-rays to smaller cryo-EM facilities, structure determination may be outsourced to external facilities like DNA sequencing nowadays, and sets of protein structures for entire cells or organisms may enable novel *in silico* approaches to understand the dynamics of life, similar to what genetics has seen in the past decades.[267] This is all speculation — however, these musings, as much as the developments of the past forty years, may attribute new layers of meaning to what was conceived as an aleatory array of words by composer John Cage, what figured as an epigraph to a cryo-EM proposal in Richard Henderson's papers almost three decades ago, and what I chose to put in front of this chapter: "Structure without life is dead. But Life without structure is un-seen."[268]

Conclusion–from Stoff *to molecular pump*

In March 1976, UCSF invited journalists to a press conference at the Parnassus campus. The topic was the "new bacterial system" discovered by UCSF scientists that, as the journalistic formulation promised, would convert "sunlight into chemical energy and food."[269] The meeting involved Stoeckenius, his lab members Roberto Bogomolni and Richard Lozier, as well as Janos Lanyi from NASA's exobiology program at Ames. A press photo shows Lozier and Stoeckenius presenting an image of the Cambridge model to the audience. The take-home message was that BR represented a pump able to convert solar into electrical energy in biological cells. Insights into basic biological processes were expected, but the ruling theme of the announcement was clearly future energy technologies. In the days of the first oil shock, the press would subsequently disseminate the Californian vision of the purple membrane as a "power source" of biological solar cells and a surmised "scientific goldmine" far beyond the US.[270] The press conference and its medial precipitate is important not only because it showed the stellar ascent of a research object that Delbrück had called the "weird purple membrane" only five years previously, but also because it illustrates how the material substance rapidly turned into a crystallization nucleus for molecular biotechnologies. One may suspect a mere slip of language or a bold extrapolation by journalists, scientists, or both, but the arguments put forward in the press performed a leap from a molecular pump found in the cell to pumping, that is, energy-generating devices as working technologies. In the particular moment of the mid-1970s, a molecular machine such as this lent itself to speculations and visions that foreshadow 1980s nanotech, but that also remain distinct from it (see chapter 4). The public attention that the new molecule received was also the moment when a linearized version of BR's discovery story became known, e.g., from an article in *Scientific American*, which omitted the complete turn of the project in early 1970, and played down Blaurock's and Oesterhelt's contributions.[271]

Thus, the 1976 press conference marks an end point to the historical development described in this chapter: The materialization of a membrane pump, from the find of an active material substance, or Stoff, to the isolation of a molecule, to the characterization of its effects as a number of steps in a mechanism and finally to a visualization of its structure. This process involved a transformation of the project originally conceived at Rockefeller from membrane to molecule, the recruitment of various experimental technologies and actors, as well as the involvement of research fields and institutions from different countries. Notably, these included historically better-known, cutting-edge endeavors such as bioenergetics and structural biology, but also ones that have received less attention, such as physiology or organic chemistry. Within a little more than five years, a novel research field coalesced around the material opportunity provided by the purple membrane, which, in turn, helped move membranes from elusive, hard-to-grasp biological entities toward tangible substances. Researchers had something in hand to toy around with and to observe with their eyes. In brief, they disposed of membranes as Stoff in an almost phenomenological sense. This Stoff could be used in simple assays, tried out with new instruments, or analyzed chemically. There have been other moments of "materialization" in membrane research; yet, it is safe to say that hardly any of these objects promised to be so easily accessible and suitable to work with. BR indeed represented a "very well-behaved" research object, allowing scientists to carry out projects that would not have been possible otherwise.[272]

The developments described in this chapter coincided with big changes in membrane research, as epitomized in the fluid mosaic model from 1972. This latter quickly displaced the old Davson-Danielli structure and suggested that biological membranes be thought of as an arrangement of two lipid films with proteins sitting therein and traversing them from face to face, as not only Henderson and Unwin's model had confirmed. Moreover, the films were imagined to exist in a dynamic, liquid crystalline state, with the proteins "drifting" in a moving sea of lipids. This model, as many other findings emerging from, for example, immunology or studies of erythrocyte membranes, ended old debates and opened new research questions.[273] And even if the history of this rapid development toward a molecular view of and approach to membranes would look different if different case histories were considered, these would presumably converge on similar molecular-mechanical models to explain function—with "hinges" or "gates" regulating the opening and closing of channels, "twists" of the protein transmitting signals across the membrane, or "tilting" protein segments transporting substrates from the inside to the outside.[274] The materialization of pumps, channels, and membranes that started in the 1960s, and that gained full force in the period described in this chapter, shows some parallels to the materialization and molecularization of the gene concept in the 1950s, with DNA or the respective proteins becoming stable but still flexible objects re-grouping researchers, instruments, expertise, and problems.[275] In the present case, this would lead to the formation and unfolding of a broader molecular-mechanical vision of life, as described in the next chapter.

In contrast to, for example, developmental biology or genetics of higher organisms since the 1970s, the membrane and protein field retained molecular biology's heritage of reductionist analysis, modeling, and thinking in quite an extreme form. Not only membrane proteins, but also other active substances such as enzymes were increasingly conceived of as robot-like contrivances of nature to carry out biological function on the molecular level, so that biochemist Arthur Kornberg would later on simply equate enzymes and molecular machines, as in the epigraph to the Introduction.[276] The models and images of such machinery and its workings do not convey its materiality very well, or even render it as somewhat immaterial. By contrast, this chapter has shown that the molecular machinery that surrounds us today has emerged from a laboratory world of working with active material substances, or the Stoff of life— by isolating fractions in the centrifuge, by observing their colors and reactions, by extracting and mixing substances, by shooting electrons at them or creating models of them. The materiality of life mattered for the shaping of the concept of a molecular machine, and it still does.

Part Two
Remaking Membranes and Molecular Machines

3

Synthesizing Cells and Molecules—Mechanisms as "Plug-and-Play"

However, the idea of life out of an automat is "nonsense of course," as chemist Thomas Dörper suggests. The gene machine's operator knows: "Pure chemicals enter the machine and a pure product of chemical synthesis comes out of it. Chemically considered, it is DNA, but biologically considered, it is absolutely dead."

German newspaper *Die Zeit*, 1983[277]

One way of examining how researchers would nowadays probe the function of a membrane molecular machine would look like this: Take a vial with the pump molecules, produced by genetically engineered cells and purified, tag them with chemical probes, integrate them into synthesized model membranes, and record how the protein changes its shape upon the addition of substrates.

From this short description, it is clear that all of the mentioned components of life—protein, membrane, tags, and substrates—represent made-up and pieced together material substances. That is, researchers probing molecular mechanisms of, e.g., pharmaceuticals work within an arsenal of living substances that is synthetic in many ways: Produced by genetically engineered organisms or made by automats assembling molecules, modified by attaching tags or probes to it, and finally assembled into a cell-like structure that can be researched in a "plug-and-play" mode. In other words, the materiality of life in contemporary science is strikingly different with respect to the "building blocks" or compounds and their availability as compared to forty or fifty years ago. This chapter will detail how we got to this point, thereby focusing on the relevance of ways of making, or syntheses, for a molecular-mechanical understanding of life.

The interplay between the making of an object and the understanding of that object is nowhere in science as important as in chemistry. Since the nineteenth century, when organic chemists produced a plethora of novel substances in their retorts, syntheses have served to create substances for many purposes—to make novel artificial materials from scratch (synthetics), to imitate natural substances for use (dyes), or to remake natural compounds in order to confirm knowledge of their chemical structure and composition (complex organic, i.e., carbon-based substances).[278] Concerning the entanglement of making and knowing, Catherine M. Jackson has convincingly argued that the origin of organic synthesis in the work of nineteenth-century German chemist August Wilhelm Hofmann, which was later overshadowed by the commercial success of constructive chemistry, was more investigative than economically motivated, centering on reactions rather than product.[279]

As much as syntheses have served different purposes, the meanings of this concept have been manifold, referring to futurism as well as to the (non)genuineness or man-made character of a product, or to technical routines to make it.[280] It is the latter two aspects that are central as to how the impact of synthetic practices on the life sciences are questioned in this chapter. By analyzing how humans have technologically remade components of organisms, I take up the obvious, but still lingering question (hinted at by the epigraph) about the relationship between chemical syntheses and ways of making in the life sciences, especially but not exclusively with regard to recent synthetic biology. Even if a plethora of research endeavors to design, engineer, or remake organisms or parts of them have been baptized "synthetic biology" throughout the twentieth century, and the current field can be considered neither their heir, nor simply a "replica" of nineteenth-century synthetic chemistry, as Luis Campos and Bernadette Bensaude-Vincent have convincingly argued, the relationship between today's synthetic biology and synthetic chemistry especially on the level of practices needs to be questioned: How have these fields been related historically as syntheses of biological substances—from molecules to cell-like aggregates—become ever more influential and widespread?[281] Have the recent life sciences indeed been facing another round of natural objects turning synthetic—whereas this applied to carbon-based organic chemical substances before, it is now about the substances and components of life? What were the epistemic, practical, and disciplinary relationships between organic chemistry and the life sciences in the late twentieth century? How far have the image and the success of synthetic chemistry actually informed the dealings with life one hundred years later?

This chapter suggests an answer to these questions not by looking for precursors of synthetic biology in chemistry, but by taking into view a *genealogy of chemical practices* from the 1970s and 1980s, which have formed a basis for contemporary synthetic biology at the interface of the chemical and the life sciences. This pertains to the adoption of technique and concepts from chemistry into the life sciences —such as "total synthesis," or making from scratch—as well as to the adoption of instruments and machines to take apart and remake the complex substances of life.[282]

Most importantly, this chapter focuses on the targets of this research, that is, the materiality or the Stoff that life is made up of, and the interplay between gaining chemical knowledge of its composition and dynamics, and the development of ways to (re)make it. Practices of analysis and synthesis, of taking apart and putting together DNA, proteins and cell-like aggregates, which fall within both the realms of chemical and molecular biological thinking and working, have significantly changed the entire material inventory of life after 1970. As a result, functional biological matter was rendered much more akin to substances in test tubes (and analogous to man-made machinery) that can be de- and recomposed or modified. This transformed materiality of biological matter has contributed significantly to the shaping of the contemporary molecular-mechanical vision of on protein function: Life has become mechanical, or so I argue in this chapter, because it can be taken apart and put back together on the level of molecules and even cells.

I will venture into the as yet largely uncharted historical territory of the intersection between organic chemistry and the molecular life sciences in the last quarter of the twentieth century, which includes figures such as the biophysicist Alec Douglas Bangham and biochemist Efraim Racker or the organic chemists Har Gobind Khorana and Bruce Merrifield. Thereby, this chapter brings to the fore the important and frequently underestimated share that chemical practices to make things have had on the formation of the present molecular life sciences.

As in the preceding chapter, my analysis will continue to follow the case story of bacteriorhodopsin, as it comprises a number of important developments that have also become influential on a more general level; however, I will zoom out of the case study at times to describe a larger picture of my actor's research or the introduction of new methodology.

An appropriate point of departure to conceive of this episode from membranes to molecular machines is the 1976 San Francisco press conference on BR: Even if quite a lot was already known about the molecular structure and mechanism of the pump as compared to other proteins, this and most other such biological substances differed from other organic (i.e., carbon-based) chemicals in an important way: The pumping molecule was still a *natural product*, assembled by living cells and to be isolated from them by biochemical techniques. This was to change in the period until 1990, when scientists developed various strategies of taking apart and remaking proteins, based on techniques from organic chemistry as much as from the newly developing recombinant DNA. Even if many limitations existed in practice, it had become possible at the end of this development to make and modify a molecular machine and put it to work in a cell-like structure in ways that I will detail below. Insofar, this chapter can also be framed as the transformation of life's material inventory from *Stoff of nature* toward *Stoff of the laboratory* (i.e., man-made, mobile, controllable, modifiable).

In spite of all particularities of a case story with respect to the timing and sequence of events, I argue that this transformation represents a general trend and has become a hallmark of the present molecular life sciences. Research on life is carried out widely in the "plug-and-play" mode sketched above. Insofar, this story should also be understood as explaining how biological processes became reformulated as molecular-mechanical processes through conceptual and methodical transfers and exchanges with organic chemistry, by piecemeal additions of technique and instruments such as automats to make DNA and proteins. This has rendered the boundary between chemical substances and biological matter permeable, and in turn also that between the chemical and the life sciences.

Making cell simulacra in the test tube—liposomes

"Six lyophilized egg yolks are homogenized with 150 ml acetone in a top-drive macerator, allowed to stand for 15 min, then centrifuged at 400 x *g* for 15 min to remove acetone-soluble impurities (triglyceride fat, steroids, and pigments)." Thus goes a recipe, published by Alec Douglas Bangham, originally a hematologist, in 1974, that had nothing to do with food chemistry, but was, in fact, the first step toward making material models of cells in a test tube. And here's the result: "If now the smear is gently covered with a thin layer of a dilute salt solution, say, 10 mM, smectic mesophases will rapidly form, and the observer will be beguiled by the process." (fig. 14)[283]

In fact, Bangham was neither the only scientist, nor the first, to be beguiled by smectic mesophases (a term from crystallography), or in other words, self-assembling vesicles enveloped by a double lipid membrane. These little spherical blobs, later called liposomes, not only resembled cells under the microscope, but allowed many of their physiological phenomena to be reproduced, thereby becoming mysteriously lifelike objects in the test tube. Historically, strategies to obtain insight into membrane or cell structure and function by re-forming films or liposomes (with lipids isolated from actual biological substances) hark back to interwar colloidal and surface chemistry. Similar models of cell and membranes had been prominent with, for example, British biologist William Bate Hardy, chemist Eric Rideal, or American physicist Irving Langmuir.[284]

Whereas Bangham called his liposomes "surrogate cells," as they were in some sense a replacement of the actual objects, I will designate them *cell simulacra*, as their function in research was not so much to be a material ersatz of cells, but a re-formed object displaying important structural and functional properties of the living cell. [285] Without buying wholesale into the philosophy of Roland Barthes here, it is notable that the characteristics and function of liposomes exemplify what he wrote at the time of Bangham's experiments about the production of simulacra in the context of the "structuralist activity:" The aim of the latter was to reconstruct an object in order to understand its rules of functioning, thereby creating a "directed, *interested* simulacrum" that is not an "original 'impression' of the world, but a veritable fabrication of a world which resembles the primary one, not in order to copy it but to render it intelligible."[286] It is along these lines that liposomes, resulting from a disassembly of an original structure (lipid extraction from a cell) can be understood as reformed simulacra, meant to represent properties of living cells and membranes in a specific situation and for a given moment. Thereby, they become an instrument of explanation, or, to bring Barthes into conversation with his unlikely kindred spirits from biochemistry, a material *explanans* of the living *explanandum*. Addressing the aspect of uncanniness that comes with man-made look-alikes of organisms, physiologists have fittingly called such cell simulacra "ghosts."[287]

To understand what was meant by reconstitution, let us look at a short note published in *Nature* in 1962, *announcing* a "reconstitution of cell membrane structure in vitro and its transformation into an excitable system."[288] What was described in this paper, the formation of so-called black lipid films, seems to have been exciting indeed: "The

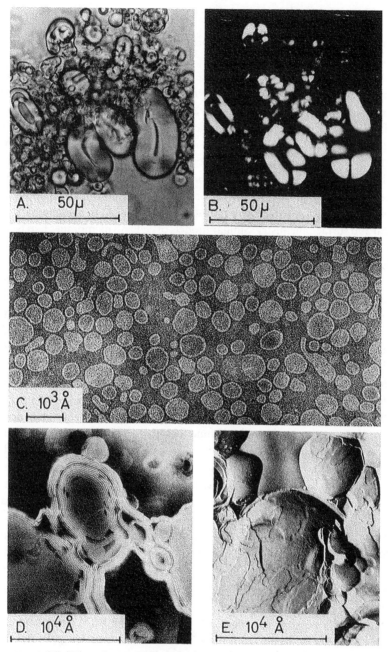

FIGURE 14 Cell simulacra. Images of lipid vesicles, i.e., membrane-enclosed droplets made from, for example, egg yolk lipids (lecithin) immersed in watery solutions. Similar vesicles were used in interwar colloid chemistry as material cell models. Under the name "liposomes," they saw a revival in the age of molecular biology and bioenergetics since the mid-1960s, and were later recruited for the delivery of drugs in biomedicine or to the cosmetic industry. Top row, light microscopic images, middle and lower row, electron micrographs of stained liposome preparations. Image E has been obtained by freeze-fracturing the sample, i.e., by mechanically breaking up the membrane layer after flash freezing, which reveals its different layers. From Bangham, Alec D., M. W. Hill, and N. G. A. Miller. 1974. "Preparation and Use of Liposomes as Models of Biological Membranes." In *Methods in Membrane Biology*, edited by Edward D. Korn. New York: Springer, p. 4. Reproduced with permission.

physiologists went mad over the model" that was "as irresistible to play with as soap bubbles," remembered liposome enthusiast Bangham, but they were not the only ones.[289] Max Delbrück was quick to congratulate the authors of the work, and adopted black lipid films in his Caltech lab for the molecular study of sensory perception (albeit, as in the case of BR before, with little success).[290] But what exactly was described in the note? Similar to Bangham's protocol, the authors of the *Nature* work had chemically extracted membrane lipids from cell material, and used these to re-form a thin lipid film at an interface separating two liquid-filled compartments. In other words, they had mimicked the elusive cell membrane in a laboratory setting by reassembling it from its components, such as lipids isolated from egg yolk or soy beans. Remember that the assembly of small lipid molecules into a colloidal vesicle or sphere is a spontaneous, self-organizing process in a watery solution. As the electrical behavior of this membrane simulacrum appeared similar to models of electrophysiology—it could be electrically charged and discharged in a similar way to nerve cells—the authors concluded that they had *reconstituted* the cell membrane structurally and possibly also functionally.[291] Thus, it was the lifelike behavior of black lipid films or vesicles that garnered attention, and reconstituted membranes of various sorts developed into models for physiologists, pharmacologists, cell biologists, and many others.

Liposomes, coming even closer to cells than films, were used to study phenomena from self-organization and the origin of life to membrane processes in bioenergetics or cellular signal transduction. They have also attracted the attention of medicine and the cosmetic industry: Bangham remembered not only scientists, but also L'Oréal and Christian Dior visiting his institute, who were interested in novel ways to deliver and distribute substances to surfaces such as the skin. As a consequence, the term liposome, as much as micellae (designating similar vesicles formed by a single layer only), has become part of advertising language. From targeted drug delivery to food production, liposomes function as "Trojan horses" to transport substances into cells, using their membranes as a cover, and they have nowadays become common stock in science, biomedicine, and consumer culture.[292]

Reconstituting the bioenergetic cell—Efraim Racker, liposomes, and molecular machinery

Around 1970, biochemists integrated functional proteins into liposomes, thus creating "proteoliposomes" to selectively model membrane processes in the test tube and spell out the interactions and dynamics of proteins, or the mechanisms of biological processes.[293] This approach of reassembling a cell-like structure was also called *reconstitution*, and it was especially influential at the time in the study of how cells generated energy on a molecular level, or bioenergetics (see chapter 2).

Researchers who focused on the role of membrane-bound cell compartments in the generation of energy had been struggling with the fact that membranes as isolated from cells contained many different proteins, which made functional analyses on the molecular level difficult, and that, on the other hand, isolated membrane proteins could not be studied in watery solutions as could other enzymes, since they required the membrane environment and structure for function.[294] In this situation, a cell simulacrum with a known composition was considered a way out of an experimental dilemma.

A video teaching session by the British Open University from 1976 provides insight into the practices and modeling culture that accompanied the uses of liposomes in reconstitution experiments (plate 4). Setting out on an exploration of bioenergetics, the video presented a cutting-edge and all but consensual topic to students of the mid-1970s, namely the question of how biological cells generated energy in the form of ATP at the membranes of their mitochondria. What may occur on the molecular level in these cell-like organelles was explained through interviews with researchers, by means of experimental demonstrations, and with large 3D-plastic models of mitochondrial membranes and the proteins sitting therein, these latter being designated as, for example, "turbines" or other mechanical gear. Peter Mitchell's disputed chemiosmotic hypothesis of energy generation, in which membranes played a central role as a permeability barrier maintaining a gradient of protons that stored energy, was analogized to a dam that would maintain the kinetic energy of water. The presenter then announced a "fascinating experiment" in favor of this theory, which indeed became iconic for bioenergetics.

This study, carried out by Walther Stoeckenius, the co-discoverer of BR from San Francisco (see chapter 2), and Cornell biochemist Efraim Racker in 1973, was based on using the purple pump protein in a cell simulacrum to demonstrate cellular energy generation. As the video illustrates (it was shot only three years after the initial studies), the experiment quickly acquired a demonstrative function.[295] To detail the experiment, Racker appears on screen, a gray-haired gentleman in short sleeves, wearing horn-rimmed glasses and a brown tie. Racker was an Austrian émigré with a medical degree, whose research trajectory had

FIGURE 15 "Thinking with his fingers"—Efraim Racker (1913–1991) in his laboratory at Cornell University, 1966. Racker, standing in front of glass pipettes, is holding a test tube probably in order to visually assess, e.g., composition or mixing of a sample—this is frequently done in work with liposomes, and generally in wet biochemistry. In a recollection, Racker characterized his specific style of experimentation as follows: "Most of the ideas have come from experiments that I tried to interpret. I think with my fingers." (Racker and Racker 1981, 271). From: Cornell University Faculty Biographical files, #47-10-3394. Division of Rare and Manuscript Collections, Cornell University Library.

led him from brain physiology in Britain toward intermediary metabolism and later to bioenergetics in postwar New York.[296] Racker is not only the wittiest biochemist the author has ever read (and wished to have had a word with), he turned into a central player in the controversy about chemiosmosis. His laboratory not only isolated crucial protein components from mitochondrial membranes in the 1960s, among them the rotating ATP-synthase, a protein complex later to become a paradigm of a "molecular machine," it also introduced reconstitution and proteoliposomes into bioenergetics (fig. 15).[297]

Racker himself characterized this approach as an interplay between "resolving" and "reconstituting" bioenergetic membrane processes. Based on the chemical distinction between analysis and synthesis, *resolution* meant for Racker analytic biochemistry—the purification of enzymes as molecules, their functional characterization in vitro, or the profiling of metabolic activities. *Reconstitution* referred to the reassembly of the isolated components (membrane and proteins) to mimic functional states in test tube model systems of increasing complexity—such as the making of defined cell simulacra.[298]

Thereby, he took up a tradition of cell and membrane modeling that had stretched from interwar colloid chemistry to researchers such as Bangham. But whereas this synthetic approach had existed disconnected from the molecular approach of genetics or biochemistry that became cutting-edge after the war, it was to no small extent Racker who introduced the molecular dimension to reconstitution as he managed to integrate purified proteins into liposomes. Moreover, EM was used to image liposomes, and radioisotopes were added to samples to follow the ensuing biochemical reactions. For the historical development of cell and membrane biology, this meant that a tradition of modeling complex structures, and understanding by analogy, became fused with the tools and concepts of postwar enzymology and molecular biology. This combined strategy, popularized by Racker in a book under the tell-tale title *A New Look at Mechanisms in Bioenergetics*, had a major impact on the development of the field.[299] Reporting on the advance Racker's lab had made with the approach in bioenergetics, the *Miami Herald* ran an article in 1972 that the living cell's "elusive energy system" had been "partially duplicated in the laboratory."[300]

Let us return to how the 1976 teaching video introduced the BR reconstitution experiment. Racker, standing in front of his laboratory bench at Cornell, narrates with a slight German accent how "one day, Dr. Stoeckenius from California called me and told me about some experiment which he has been conducting."[301] He then presents a sausage-like plastic model of *Halobacterium*, elaborating on the purple membrane, the rhodopsin-like protein sitting therein, and the light-dependent transport of protons. Stoeckenius was invited to a seminar, or so Racker continues, and was asked him to bring along some BR, which was shown to the audience as a little flask containing a pink-colored fluid, and they set up an experiment in which the protein was incubated with lipids to form a proteoliposome. In other words, Racker and Stoeckenius joined forces to reassemble a proteoliposome containing the purple pump, and yet in other words, the new research object BR was put center stage into the bioenergetics controversy—"Protons were moved when the bacteriorhodopsin vesicles were illuminated," thus Racker on the study that was based on similar principles as Oesterhelt's pH-pulse studies described in chapter 2; however, they employed cell simulacra.[302]

In fact, correspondence confirms that the first piecing-together of a functioning cell simulacrum by Racker and Stoeckenius in June 1973 seems to have worked straightforwardly. As Racker wrote to Delbrück at the time, "I was prepared for complications . . . We were lucky, however." The incorporation of the protein into liposomes worked well, and:

> [. . .] I am just trying to combine this proton pump with the mitochondrial ATPase to get ATP generation and am getting promising results, I agree with you that this is an exciting system, but I am sure that the anti-Mitchell crowd will call it a nice model system and not accept its applicability to the natural systems.[303]

This quote shows how quickly the concept of a molecular pump was taken up by bioenergetics, and how the experimental opportunities presented by BR were adapted in a new context. However, Racker's doubts also reveal a general problem of reconstitution experiments: While allowing the complex situation of the living cell to be addressed with a model of known composition, it was problematic to assess whether the measured effects actually represented the natural situation or whether they were artifacts.

Doubts notwithstanding, Racker and Stoeckenius's study appeared in print only a good six months after the experiments were begun, in the somewhat preliminary format of a "communication" filling only two pages of the *Journal of Biological Chemistry*. The paper argued that the molecular pump, incorporated in a liposome with the other mentioned protein, i.e., the ATP-synthase of beef heart mitochondria that Racker's lab had purified, would form ATP upon illumination.[304] This study was not only frequently cited by scientists, but also discussed by philosophers and sociologists for its supposed role in resolving the controversy on how cells generate energy at their membranes. It has been designated the "capstone" of the bioenergetics controversy in favor of chemiosmosis for its implication that a proton gradient across a membrane *alone*, as produced by the pump BR, sufficed for the production of ATP by the ATP-synthase, and that *no* other components or processes were required (such as direct interactions between other proteins in the mitochondrial membrane or intermediate chemical reaction products).[305] This was basically also the interpretation proposed by Racker himself in a 1977 review, which is often quoted as having ended the controversy.[306]

However, in order to understand the role of the Racker-Stoeckenius experiment in science and historiography, one has to take into account not only its logical structure and the conclusions it permitted, but the specific makeup of the cell simulacra: These brought together an animal ATPase synthase protein and the purple membrane of a microbe in a lipid membrane environment from a plant, the soy bean. That is, molecular components from different organisms were pieced together to form a hybrid cell simulacrum.[307] This reassembly of biological function in a structured in vitro system suggested that generalizable physico-chemical processes were operative in evolutionarily distant cells, and just like parts of devices, one could combine a generic cell "casing" from one organism with components of two others to form a hybrid system in which the different functions joined together like parts of machines. The feat of putting together biological components that had no prior connection and making them work may also explain why the existing criticisms of this study did not dominate its reception.

From chemiosmosis to molecular mechanisms

What has frequently been overlooked is that the Racker-Stoeckenius experiment paved the way toward the acceptance of a specific version of chemiosmosis, namely a molecular-mechanical one, which was not the one advocated by its creator Peter Mitchell. Mitchell's correspondence with Stoeckenius from the years 1973–1979 shows that the central figure of bioenergetics became interested in the experiment, but remained somewhat cautious about the results and their interpretation. Mitchell had considered BR a "very nice system for experimental study," especially since it entered the arena as a novel object not overloaded with conflicting data as existed, e.g., for mitochondria or chloroplasts.[308] However, the very details of what Mitchell called the protein's "coupling mechanism," i.e., the way the absorption of light was coupled to the build-up of a proton gradient and the subsequent generation of ATP, remained controversial between him and the new BR researchers.

Even in letters to Stoeckenius from 1977 and 1978 Mitchell opposed the explanation of protein function by "conformational changes" (i.e., changes of protein shape, as established in the early BR work; see chapter 2), and contrasted this to his own thermodynamic explanation: This latter centered *not* on the mechanical movements of proteins sitting in a membrane, but on them catalyzing spatially ordered chemical reactions (such as the redox processes in the respiratory chain).[309] The difference between these two ways of understanding molecular phenomena is as subtle as it is fundamental: Mitchell believed that ion gradients were built up by chemical reactions oriented through a membrane, such as by having *a* proton on one side of the membrane react and thereby disappear. Stoichiometrically, this would imply the build-up of a gradient; however, there is no molecular movement of or within a molecular object required in his model! By contrast, Stoeckenius, arguing on the basis of molecular data from spectroscopy, defended in the correspondence another understanding of the same process, which implied that *a potentially identifiable* proton was pushed across the membrane mechanically by a pump without a proper chemical reaction (i.e., the breaking of a chemical bond to release or bind the proton), like the freight of an elevator. This explanation comes close to today's perspective on protein molecular mechanisms. The differences between Mitchell's concept and that of Stoeckenius, Racker and Oesterhelt, are nicely illustrated in an embarrassing conversation the latter, then a mere youngster, recalled with his senior colleague: To Mitchell's question concerning what he thought the chemical reaction (specifically, the redox reaction, the mainstay of Mitchell's explanation) behind BR function would be, Oesterhelt avowed that he supposed there would be none, as proton transport was accomplished by mechanical tilting of the molecule. Mitchell insisted that he just had not found this chemical reaction *yet*, and would not let him walk off like that.[310]

The fact that both Mitchell and Stoeckenius thought of each other's approach as a "black box" underlines that we see here not only different explanations, but what one could tentatively call, following Ludwik Fleck, the clash of two incongruent thought styles within bioenergetics, and potentially within biochemistry in general. The conflict between Mitchell's older model of explanation, inspired by thermodynamics and colloid or surface chemistry centering on chemical reactions ordered in space, and the newly emerging molecular-mechanical perspective, as epitomized by spectroscopy or reconstitution, had a paradoxical effect: Whereas Mitchell's chemiosmotic model became generally accepted in the course of the 1970s, and while he was celebrated as the "winner" of the bioenergetics controversy, the differences in the details of the chemiosmotic mechanism pitched him against the mainstream of the field, which sided for molecular mechanisms—the revolution devoured its father (for reference see note 306).

Mitchell's role in bioenergetics and his personality set aside, this refined historical interpretation of the Racker-Stoeckenius experiment complexifies the history of bioenergetics as much as it underlines the role of this study, and the used research objects, for the ascent of a mechanical understanding of molecular processes. Since the 1980s, molecular-mechanical explanations of protein function have become the dominant perspective in bioenergetics and beyond, with the proton pump as well as Racker's ATP-synthase—now conceptualized as a molecular rotor—as their poster children.[311] Structural and functional insights gleaned from other membrane proteins, such as the photosynthetic reaction center, reinforced this notion.[312] Thus, when Oesterhelt used an onomatopoeic term for a scissors' movement, of "*Schnipp-Schnapp*" (snippety-snip), to explain to the author what his molecular pump did, he pointedly grasped the mechanical essence of an understanding of biological function at the molecular level that now has become omnipresent.[313]

A plug-and-play—biology

The Racker-Stoeckenius experiment specifically, and reconstitution generally, are early examples for a style of experimenting with life that has become influential ever since. The reshaping of life's materiality as a set of components that one could take apart and put back together also had an impact on its visual representation. In the 1976 video, Racker explained the experiment with the help of material models, that is, plastic props of the molecular compounds—a large plastic ball representing the liposome, and two small ones (one white, one red) for the respective proteins (see plate 4 bottom). He illustrated the assembly, or "piecing together," of the proteins into the vesicle, by literally squeezing the red plastic ball (representing BR) into the surface of the larger vesicle, accompanied by an awkward shrieking sound. This visual, even haptic style of modeling is prominent throughout the video. It connects to the modeling culture of structural biology as exemplified in Dickerson and Geis' *The Structure and Action of Proteins*, and implies that proteins are structures similar to macroscopic material objects.[314] Racker's piecing together of the two plastic balls illustrates this thing-like, mechanical aspect of biological matter and processes in the most tangible way.

Thereby, the reconstitution experiment and its presentation exemplify what I propose to call a plug-and-play approach to life: This mechanical and manipulable understanding of biological processes and entities on the molecular level reaches from modeling to experimental assays, such as those on the color changes or the pH pulse experiments that displayed a repetitive, game-like behavior of material substances. Plug-and-play was enabled by material aspects of substances such as BR: As an ample and robust Stoff, the membrane preparation in a flask could be packed into a suitcase and flown from California to Cornell. In another lab, the substance would be mixed with molecular components of life from plants or animals and made to work with them in an assembled material cell model, in electric circuits or on microchips (chapters 2 and 4). This protein became a "molecular component" of life since it could be unplugged from its original environment, the cell membrane, and plugged into novel environment of a cell simulacrum apparently without loss of function.

Generally, resolution and reconstitution contributed to entrenching the molecular-mechanical vision of life. A combined analytic-synthetic approach has been adapted to the study of various proteins since the 1980s and 1990s, many of which have biomedical significance, such as cell receptors from immunology, ion channels of nerve, or multidrug pumps contributing to the resistance of cancer cells.[315] Reconstituted membrane proteins are even sold by biotech companies in ready-made assays to probe the function of proteins or to screen potential pharmaceutical substances. However, it should also be stressed that the plug-and-play approach that could be established so easily in the case discussed here, caused numerous problems with other biological matter, most of which was much more delicate and recalcitrant, rapidly losing its function outside of its original environment—proteins mostly did not want to play the games researchers wanted. How an exceptional case such as presented here became a pioneer for a new normal, that is, for a more widespread plug-and-play biology, will be described in the following sections.

Plug-and-play also anticipates the problem of an intrinsic modularity of life, i.e., the question whether cells harbor an inventory of molecular parts than can be isolated and recombined in order to newly design biological cells, e.g., by engineering metabolic pathways, signaling cascades, or other biological devices. Under the label of the "biobrick," such strategies have become an interest of today's synthetic biology. Simple model systems such as BR may have given rise to an impression that may not easily be generalized, and that may also have intrinsic biological limitations (regarding, e.g., delineation and combination of "parts").[316] Yet, the promise and vision of a "modular" inventory of life, to be isolated and reassembled, has remained influential in synthetic biology, biotechnology, and biomedicine. Not only have recombinant DNA techniques helped to fill laboratory freezers with protein or nucleic acid parts (as described below), nowadays, made-up cell simulacra of various sorts exist: Membrane vesicles from simple erythrocyte ghosts to much more complex artificial cells are regularly used; some of them short-lived, others longer, some for the purpose of probing cellular function in the lab, others to specifically deliver cosmetics or pharmaceuticals in the body or even to make cheese more creamy. Proteoliposomes have been used to model scenarios for the origin of cells, and thereby of life as we know it.[317] Membrane vesicles nowadays represent not only a mere casing of the "protocells" with which synthetic biologists tinker, but their active, i.e., functional boundary, mediating between an inner milieu and the environment, and thereby representing a pivotal step toward artificial cells.[318]

Seen from a historical angle on the development of the molecular life sciences, the unspectacular blobs of egg yolk lipids described by Alec Bangham at the beginning of this chapter represent a productive and influential heritage of colloidal thinking and working. Figures such as Bangham, Mitchell, or Racker, as much as membranology in general, thus defy the historiographical model of a monolithic midcentury molecularization of the life sciences along the axes of genetics and structural biology. Stories such as that of liposomes and reconstitution hark back to interwar colloid chemistry, a field that has often been declared dead prematurely after its major conceptual premises had become refuted in the 1930s.[319] The historical continuity adumbrated in this case may be described as a genealogy of practices to imitate and ultimately remake lifelike objects on the cellular level. The concept genealogy as adapted from Friedrich Nietzsche by Michel Foucault implies a historiography focusing on "those things nearest to it," and whereas Foucault speaks of the body here, materiality and practices may well take that place in the present history of science.[320] And just as Foucault highlights the "barbarous and shameful confusion" of elements found by such historiography, disconnected from an origin or a master narrative, the present case demonstrates how practices of modeling a cell-like object were successfully disconnected from the controversial conceptual framework of 1920s colloid chemistry, and that using them did not entail identifying oneself as a "colloidalist" in a sense that was tainted in the postwar molecular life sciences. By contrast, the story of resolution and reconstitution as exemplified in Racker's work illustrates how such practices of making lifelike objects were successfully combined with the molecular approach to life from protein biochemistry, and how these practices helped to pioneer a molecular- mechanical understanding as well as plug-and-play approach to life that is dominant nowadays.

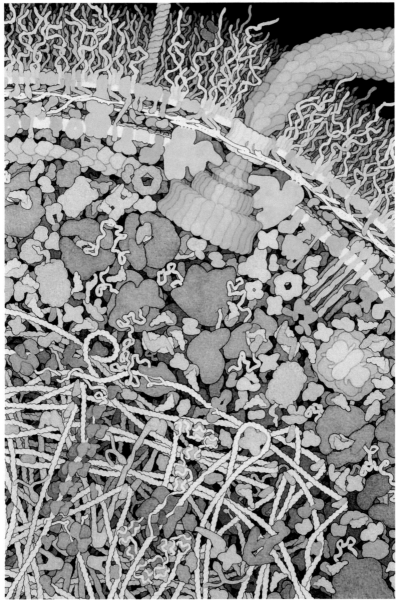

PLATE 1 The cell as a molecular hustle and bustle. Graphic based on the structural outlines of proteins, nucleic acids, polysaccharides (sugars), and lipids making up a bacterial cell. Scale is indicated as 1,000,000×. From: Goodsell, David S. 2009. *The Machinery of Life.* 2nd ed. New York: Springer, p. 52. Reproduced with permission.

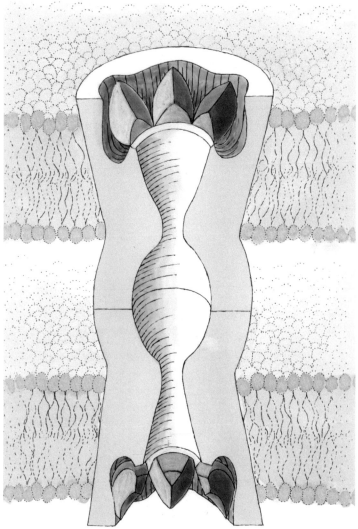

PLATE 2 Molecular imagery. Illustration of a gap junction by American artist and scientific illustrator Irving Geis (1908–1997). Gap junctions connect nerve cells (membranes of both cells depicted as yellow bilayers), and they are formed by two adjacent protein half-channels (blue). Gap junction channels (called connexons, depicted here in open state) allow the synaptic transmission of electrical signals between neurons. Although the image dates presumably from the 1980s, it is illustrative of the simplified, visually appealing style of depicting biomolecules and processes pioneered by Dickerson and Geis' textbook *The Structure and Action of Proteins* (1969). As Geis' longtime collaborator, biochemist Richard Dickerson, remembered, Geis claimed "not to draw a protein exactly as it was, but to show how it worked," to "move it out in the open where it could be seen and the molecular mechanism thereby understood" (Dickerson 1997, 2484). Similar models of macromolecules, from material 3D structures to those on screens, have informed the molecular-mechanical vision of life throughout, and become ever more widespread with the advent of computer graphics. Image from the Irving Geis Collection, Howard Hughes Medical Institute. Rights owned by HHMI. Not to be reproduced without permission.

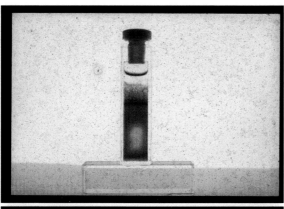

PLATE 3 "Stuff that does something." Slides used in presentations by Dieter Oesterhelt, presumably mid-1970s. Top, photo of cuvette filled with purple membrane material suspended with salt solution and ether. Bleaching from purple to yellow was achieved by illumination with a focused light beam, demonstrating the material's reactivity (photoeffect). Similar images demonstrating material activity were frequently shown in presentations by bacteriorhodopsin researchers. Bottom: Experimental set-up to demonstrate photoeffects, e.g., in lectures. Images reproduced with permission of D. Oesterhelt, Martinsried.

PLATE 4 Membrane mechanisms explained. Film stills from teaching video, 1976. Top and center, biochemist Gottfried Schatz, in front of the drawing of a mitochondrion, the opened outer membrane of which makes internal membranous invaginations (cristae) of this organelle visible. Schatz used a mobile 3D-model of a membrane bilayer with white spherical components representing the ATP-synthase to demonstrate the structures and molecules involved in cellular energy generation by a chemiosmotic mechanism. Bottom: Proteoliposome modeled with plastic balls by biochemist Efraim Racker. The small purple ball (bacteriorhodopsin) and the medium size white ball (ATP-synthase) are squeezed into the larger white ball (liposome), illustrating the reassembly of life's molecular inventory, or a plug-and-play biology. From Open University chemistry video on oxidative phosphorylation (Nunn 1976). With kind permission by the School of Life, Health and Chemical Sciences, © The Open University, http://www.open.ac.uk.

Remaking life's molecular inventory

Attempts to make not cells but much smaller biological molecules by synthesis, such as proteins, have been a mainstay of organic chemists in the first half of the twentieth century. The aim was to (re)make living substance from scratch, or, more precisely, to build large and complex biomolecules from smaller carbon compounds sitting in flasks on laboratory benches.

Syntheses of substances such as hormones (some of which were proteins or protein-like) or other complex organic compounds such as vitamins or toxins, have been considered as a chemical art, based on a skillful combination of intermediates by plentiful chemical reactions. However, the art of synthesis has served different scientific, at times very mundane, purposes: First, syntheses aimed to obtain or confirm insights into molecular structure (pertaining to, e.g., components, chemical bonds, or side groups of a large molecule) as obtained before by analytical methods, i.e., isolations or methods breaking larger molecules into pieces that could be characterized. This was often the reason for complex and expensive "total syntheses" (among organic syntheses, one can distinguish approaches to build up complex natural substances entirely from simple compounds of another source, such as carbohydrates, ammonia, or water, called total or complete synthesis, versus partial or semi-synthesis that uses larger molecular fragments). Second, achievements to remake a potent biological agent such as a hormone, minute amounts of which would transform an animal or human body, have retained a Promethean spirit of mastering nature—such syntheses were "feats" shifting the boundaries between nature and artifice, and they have certainly contributed to the cultural image of chemistry. Last, and certainly not least, the chemical remaking especially of substances such as vitamins and hormones has included endeavors to substitute natural products for nutritional or medical use with those from the laboratory, with the latter being more accessible, pure, open to modification or refinement, available in boundless quantities, and (potentially) cheaper. Such industrial processes often relied on semi-syntheses, and even if they were influential in some cases, in others they remained problematic for economic reasons as much as a perceived contrast between a natural product and its synthetic counterpart. In sum, organic syntheses up to the first half of the twentieth century, situated at the crossroads of pharmacology as well as organic and biological chemistry, have included epistemic aspects (insight into structure of unknown biomolecules, such in the case of plant alkaloids) as well as demonstrative aspects (proof of principle, "feat," such as in the cases of vitamin B_{12} or a gene, as discussed below) as well as economic or technological aspects (such as in the case of vitamin C, a consumer product synthesized by combining chemical and biotechnological steps).[321]

Surprisingly, the impact of synthesis, and generally that of organic chemistry, on the development of the molecular life sciences after World War II has garnered only very limited historical attention. This is a desideratum of research since this period saw widely perceived syntheses of biomolecules. Just to name a few, syntheses of various hormones were achieved, of the photosynthetic pigment chlorophyll by Robert B. Woodward of Harvard University, or of vitamin B_{12}, jointly achieved by Woodward and Albert Eschenmoser from the ETH Zurich in 1972.[322] In this section, I will describe the impact of syntheses on the molecular life sciences by looking at novel ways to make ever larger biological molecules from scratch, which culminated in the creation of functional synthetic genes and proteins in the 1970s and 1980s. One central protagonist of this story will be the Indian-American organic chemist and molecular biologist Har Gobind Khorana. I will also glimpse into the mechanization of protein and nucleic acid–making through machines such as by Robert Bruce Merrifield's peptide synthesizer. My central argument will be that with these chemical methods, in combination with the establishment of recombinant DNA and biotechnologies in the 1980s (a much better studied subject), biological macromolecules increasingly changed from *products of nature*, isolated biochemically as the cells made them, into *products of the chemical laboratory*, that is, pure and in many ways synthetic substances—characterized down to the atomic details of their composition and structure, (re)made, modified, taken to pieces and reassembled or even designed. Novel ways of making molecules at the interstice of chemistry and the life sciences, or so I will argue, steered the materiality of biological objects significantly toward the state we now encounter in the sciences and technologies of life, and this should be seen as another important step that contributed to turning molecules into machine-like objects.

Synthetic molecular biologists—making molecules in retorts and by machines

Among the well-known figures of molecular biology such as Crick, Delbrück, Jacob, Monod, or Watson, all of which were either molecular geneticists or structural biologists, the personality and work of Har Gobind Khorana (1922–2011) has garnered curiously little attention.

Even in histories of the genetic code, for the deciphering of which Khorana shared a Nobel prize with Robert W. Holley and Marshall W. Nirenberg in 1968, his contributions, centering on the use of synthetic DNA messages to record the "answer" of the cellular apparatus to genetic messages, have not been discussed in great detail, which stands in stark contrast when taking into view his accomplishments and the impact he made on the field.[323] This neglect may be due to the fact that he never became a public figure and that he did not indulge in autobiographical or popular texts; however, what may be a more relevant reason is the fact that he pursued an approach to molecular biology that does not fit with the existing narratives of this field as a combination of molecular genetics and structural biology (i.e., DNA and protein crystallography).[324] Khorana, by contrast, pursued a decisively chemical approach to molecular biology, in which syntheses played a central role (fig. 16).

Obituaries remember Khorana as a quiet, almost humble man and an arduous laboratory worker.[325] After studying chemistry from 1940 at the University of Lahore, then part of British India, he went to Cambridge to conduct doctoral work on the chemistry of proteins with Alexander Todd, more specifically on the peptide bonds linking the amino acids of a protein's polypeptide chain. Following the war, and personal turmoil after the split between India and Pakistan, he went to the ETH Zurich's organic chemistry department for a postdoctoral sojourn, stumbling there, or so goes his own account, upon forgotten chemical reagents called carbodiimides, which allowed him to embark on a project that was to characterize large stretches of his scientific life: A way to build up ever larger molecules, or more technically speaking the synthesis of molecules of life from ordinary chemicals used as building blocks.[326] Among the molecular inventory of life that Khorana reproduced in the laboratory was, first, ATP, the central energy metabolite of the cell, and a focus of attention in intermediary metabolism biochemistry or bioenergetics (see chapters 1 and 2). As ATP contained a nucleotide ring structure as its basis (i.e., adenosine, a complex cyclical carbon and nitrogen structure that also forms one of the components of DNA), he was able to adapt his technique to make oligonucleotides, i.e., short polymers of the nucleotide bases. Such snippets of the hereditary substance had moved into the center of molecular biology since Watson and Crick's model thereof. The next step in this synthetic chain was an entire co-enzyme, still based on the same nucleotide core structure, but larger and more complex. Interestingly, Khorana's approach does not fit within the frameworks of either molecular genetics or metabolic biochemistry of the time, but bridged these by the molecules

FIGURE 16 Chemical synthetic biologists. Har Gobind Khorana (1922–2011) and a group member identified as T. Mathai Jacob in the laboratory at the University of Wisconsin, operating a chromatography column. The photograph was taken in the 1960s, presumably in the context of Khorana's work on the genetic code or the synthesis of a gene. Chromatography columns are used widely by biochemists and organic chemists to separate molecules differing in molecular properties, such as cell contents or products of syntheses. From University of Wisconsin-Madison Archives, Image ID S10961. Reproduced with permission.

he made. His systematic, bottom-up approach to rebuild molecules, the biological assembly and function of which others studied, appears as that of a construction worker and architect, stepping on each stone he laid to reach even higher, and using earlier results of syntheses as starting materials for the next step. Khorana's synthetic chain shows some resemblance to what nineteenth century chemist Marcellin Berthelot had demanded for a synthesis of living matter—advancing from the simple to more complex, step by step in a rational way, building each new level on top of the previous.[327]

Making ever larger and more complex molecules, Khorana criss-crossed between different fields. His strategy was not an invasion of molecular and cellular biology by organic chemistry, but looks more like an integration of them, for example when he interlaced chemical methods of synthesis in test tubes with enzymological reactions, that is, the ways by which cells assembled biomolecules.[328] In these cases, it was Khorana, the chemist, who was to learn from organisms to imitate biological reactions, and to move away from the harsh and aggressive conditions his profession typically employed in their retorts, which did not work well with many biological materials. An example of such a "biomimetic chemistry" was the use he made, in a collaboration with biochemist Arthur Kornberg, of snake venom as a type of molecular "scissors" to specifically split nucleotides.[329] Kornberg, remember the epigraph to chapter 2, was an expert on the enzymes that made the DNA in cells, seen as early examples of molecular machines.

In 1960, Khorana announced the "total chemical synthesis of macromolecules possessing biological function" as a challenge of organic chemistry.[330] Thereby, he meant making a gene from scratch, and this was to become a project that would keep him and his large research group busy for the next 15 years. By the mid-1970s, when Khorana ran a laboratory at MIT, this goal had become a reality: To make not only DNA, but a functional gene from simple carbon, nitrogen, and phosphorus-containing reagents, his laboratory had developed a long sequence of synthetic reactions, blending organic chemical and enzymatic steps, and employing cutting-edge physical detection methods such as gel electrophoresis or mass spectroscopy to analyze the reactions' intermediates and products. Over the course of more than a decade, the numerous steps of the "epic task" to make a gene, as the New Scientist put it, precipitated a series of technical, tedious-looking papers authored by the teams carrying out the delicate wet chemistry of making, joining, or cutting molecular fragments.[331]

These papers also highlight that behind the Promethean façade of projects such as this one—remaking the material basis of heredity—syntheses often were sophisticated but mostly empirical and laborious collective projects. What is more, in this case, as in others, the published procedure to make a gene never became a practical routine, and a journalist rightly suspected in 1976 that Khorana's synthetic gene would be the first and last of its kind, as the "more glamorous" and "far less time-consuming" techniques of recombinant DNA that had been established in the meantime allowed some of the chemical methods to be dispensed with.[332] However, a gene as a complex DNA molecule was certainly no ordinary stuff. In addition to the demonstrative accomplishment of making what many considered as life's master molecule, this feat could be seen as opening the door toward the modification or the remaking of genes.

Khorana was not the only chemist venturing at the time into syntheses of complex and important biological substances. Another example of a chemical molecular biologist was Rockefeller University chemist Bruce Merrifield, who took up the much debated and tried-out project to synthesize proteins in test tubes. Whereas peptides (i.e., small proteins) such as the hormone insulin had already been produced in this way by the 1950s, Merrifield succeeded in piecing together an enzyme composed of more than 100 amino acid components in 1969. His remake of ribonuclease A, as this well-studied example of a nucleic acid-splitting enzyme was called (which was commercially available, and its molecular structure known), showed biological function in a test tube assay. Again, this should be seen an accomplishment if not to make a "dead" chemical alive, then to emulate some of its specific vital functions, thereby subverting the gap between the living and nonliving. Both for the gene and the enzyme, syntheses showed science's power and promise to question ontological divisions, and so these projects were much more Promethean than technical.

Even if neither Merrifield's nor Khorana's syntheses of genes or proteins became used on a larger scale (they were indeed surpassed by biotechnological methods after 1970), both have contributed in the longer run to the establishment of automated synthesis robots producing these substances as probes and reagents for molecular biology.[333] Merrifield had already constructed such machines turning the sequential synthesis of a protein's polypeptide chain into an automated process since the 1960s (fig. 17).

Basically, his apparatus consisted of a reaction vessel connected by tubes to a number of reservoirs containing the specific amino acid components and the required reagents to be added. The trick he employed was to perform the synthesis reaction not in a liquid medium as known from organic chemistry, but to immobilize the growing peptide chain on a solid support, such as polystyrol beads, in the reaction vessel. This would prevent unwanted side reactions of the nascent molecule, which would inevitably occur in solution, and allow the remaining reagents to be washed off by a mechanical pump before the next reagent was added. A similar strategy for an automated "solid-phase synthesis" has been established for nucleotides around 1980, using not exactly Khorana's

FIGURE 17 Making molecules with machines. Robert Bruce Merrifield's automated peptide synthesizer, used at Rockefeller University since the mid-1960s, on display at the museum of the Science History Institute, Philadelphia. The mechanical drum on the right executed a program, according to which motorized syringes added reagents (e.g., amino acids, solvents) to a reaction vessel (small vessel right of the big brown flasks). This vessel contained a solid-phase support material attached to which peptides were synthesized. In one step, an amino acid was added, reacting with the support material, then the material was washed before the next amino acid followed, etc., until a peptide chain had formed, which could then be cleaved off the support. Reproduced with kind permission of the Science History Institute, Philadelphia, PA.

protocols, but similar schemes to obtain a chain from specific nucleotide building blocks. Such "gene machines" made quite a splash in biotech and academic research, as they rapidly supplied ever cheaper specific DNA probes required for genetic engineering, PCR, and sequencing. Today, the research supply industry delivers entire genes synthesized on the basis of these methods at a mouse click.[334] That is, the impact of syntheses as practiced by chemical molecular biologists such as Khorana or Merrifield on today's molecular life sciences can be seen both in rendering biological macromolecules akin to more ordinary stuff that can be made and remade by humans, and more practically, in helping establish methods and approaches to mechanize these processes, enhancing speed, precision, and automation of syntheses.

Making and unmaking molecules for structure and mechanisms

With the molecular mechanisms of heredity (seemingly) left to be ironed out, many established molecular biologists were looking for new topics in the early to mid-1970s. Among target areas of this "mass migration" ranged biotechnology and biomedicine, the biology of higher organisms, and thereby development, or cellular communication, cognition and bioenergetics, the latter pivoting around membranes.[335] Following Delbrück, Khorana moved to membranes when quitting molecular genetics, considering this later as an entry point to "molecular neurobiology and signal transduction."[336]

To him as to many others we have heard of before, the purple pump appeared as a material opportunity to enter the membrane field, and he gathered expertise from the small BR community of the mid-1970s, encompassing Henderson, Racker, and notably Stoeckenius, in "getting the system going," as he put it.[337]

Khorana's first "go" at BR was, in fact, not synthetic, but analytic. His group set out on the task of disassembling the protein into its component parts, amino acids, or to be more precise, to sequence the protein. Protein sequencing was routine by the 1970s. Since the pioneering work of Frederick Sanger on insulin in the 1950s, sequencing by sequential degradation of a protein's amino acid chain was widespread, and automated apparatuses existed.[338] However, membrane proteins had proved recalcitrant to the customary approach for the reason that they were insoluble in the watery solutions used as reaction media. Here, Khorana took a novel approach by adapting one of the new physical methods of chemistry, mass spectroscopy, to the problem.[339] While his group dissected the protein into protein fragments (peptides), a mass spectrometry group at MIT determined these fragments' fine structure. This combined wet chemical and automated physical approach led to the sequencing of BR.

However, by the time the American sequence of the protein's amino acids was published in 1979, a map depicting the protein as a chain of amino acids, a Soviet sequence was already in existence, worked out by the group of organic chemist Yuri Ovchinnikov at Moscow's Shemyakin Institute of Bioorganic Chemistry.[340]

What did these hand-drawn maps of a protein as a sequence of amino acids contribute to an understanding of its molecular mechanism? Certainly, they did not simply reveal "parts" of the pump in the sense of pistons or levers. However, this map allowed understanding how the three-dimensional protein molecule known since Henderson and Unwin's 1975 EM-structure was built up from its elements, the amino acids, and, conversely, how these were related to components such as the alpha-helices traversing the membrane. It was now possible to describe the "molecular landscape" of the pump sitting in the membrane in greater detail, and to question how the emerging elements of protein substructure (i.e., protein domains, or, even smaller molecular components such as single important amino acids and their chemical groups) could be related to function in space and time. In sum, this map contributed to understanding how suspected functional elements of a molecule orchestrated its movements or shape changes. Such a presumed analytic relationship of whole and parts in terms of assembly and function is characteristic of mechanical contrivances.

In that sense, the map of bacteriorhodopsin's "secondary structure" (the technical term for the amino acid sequence of a protein, as opposed to its tertiary, spatial structure) could also be read as an instruction for further experiments to relate the protein's function to the role played by its components, that is, to determine which of its amino acids represented which functional part. A similarly consequential map of BR followed in 1981, when the sequence of the gene corresponding to the pump protein was published.[341] Therefore, Khorana's group had transferred their expertise in nucleic acid chemistry to the new membrane project: Amidst the first successful gene clonings for proteins and their sequencings around 1980, they had provided a genetic map and a DNA sequence that was the instruction for the cellular synthesis of the pump molecule.[342]

The rationale behind Khorana's introduction of molecular genetic routines into the membrane project, however, was not primarily to understand a genetic problem. By contrast, "fishing" the BR gene and sequencing it allowed recombinant DNA methods to be applied as a *toolbox* to scrutinize protein function in more detail. Remember that BR, and in fact much of membrane biochemistry and biophysics, had been a "molecular biology without DNA" before: Protagonists of this book such as Henderson or Oesterhelt had carried out their respective work mostly without addressing the level of DNA, and as late as 1976, a paper on the "biosynthesis" of the purple membrane did so entirely from a biochemical vantage point, neither wasting a word on the potential genetic regulation of the process, nor employing terms such as "operon," "transcription," or "translation" that were omnipresent in the contemporary discourse of molecular genetics.[343] This underlines the hiatus that existed prior to 1980 between biochemistry/biophysics versus molecular genetics.

A recombinant DNA toolbox such as used by Khorana and others for protein function studies became publicized from the early 1980s by, for example, Tom Maniatis's "cookbook" *Molecular Cloning: A Laboratory Manual*.[344] This indicated a shift from gene cloning as a research project to its employment as a technical routine to enable other experimental manipulations: Genetically engineered cells became tools with which to synthesize, modify, and study proteins, adding to the battery of organic chemical methods to make and modify biomolecules.

Molecular infrastructures—convenience genes

The work of Khorana's group is a good example of how analytic and synthetic approaches to biomolecules went hand in hand; that is, how the making of genes and proteins, as well as their redesign, was related to studying their molecular structure and function. To this purpose, the group created an infrastructure of recombinant DNA methods and tools, which will be explained at the example of Khorana's work on the BR gene. The accomplishment in this case was not in synthesizing this specific gene, as it had been in the 1970s; this was now easily achieved by recombinant DNA methods.[345] The aim was to include the stretch of DNA coding for BR in a genetic "cassette," thereby creating a "convenience gene." Philosopher of science Ulrich Krohs (2012) has coined the term "convenience experimentation," referring to semi-automated apparatuses and ready-made experimental kits used in the molecular life sciences of the 2000s, which standardize and black-box parts of the experimental work, thereby facilitating it. The design of genes suitable for manipulation and their distribution in the community should be seen as an early step of such developments, although not yet on the level of commercialized products. Khorana's convenience gene could easily be taken in and out of its cassette by means of restriction enzymes (i.e., the molecular scissors of recombinant DNA, cutting the strand at specific sites); algorithms were employed to determine a sequence of the DNA stretch that was more user-friendly for further cutting and pasting.[346] The convenience gene could be spliced, transferred, and distributed more easily, forming part of an infrastructure to use recombinant DNA for protein studies. It allowed one, for example, to introduce the gene into another organism such as *Escherichia coli*, a model of molecular genetics and workhorse of genetic engineering. This made the DNA more easily accessible to manipulations, many of which were difficult to adapt to organisms with an exotic physiology such as *Halobacterium*.[347]

The creation of a molecular genetic infrastructure for protein studies was in itself research work and kept Khorana's group busy for years. Yet such an infrastructure, at the time established for other proteins as well, enabled many of the later structure-function studies (such as by mutation analysis). In addition to the convenience gene, this infrastructure comprised, e.g., genetic vectors such as plasmids, cloned and/or mutated variants of the gene, recombinant organisms making the protein ("expression systems"), and assays to check function. The aim of Khorana's work, oscillating between recombinant DNA and chemistry, was perhaps not so dissimilar to that of German chemist August Wilhelm Hofmann's in organic synthesis a century earlier: In both cases, it was not primarily about producing novel substances for use, but about establishing routines to make molecules and to take them apart, which helped to scrutinize molecular structure and reactions in greater detail.[348]

With respect to the central theme of this book, the materialization of molecular machines, the existence of this infrastructure brought in significant changes, for example, when Khorana used it for the "total synthesis" of an entire mammalian receptor protein in a plant cell extract (if also in minute amounts only).[349] Not only did the meaning of the term synthesis move here from chemical to molecular biological making, this hybrid test tube system to make protein exemplifies the changes to "what it was to be a protein" after 1980, as compared to the proteins as natural products that Oesterhelt and Henderson had toiled with a mere decade before. "Protein engineering" transformed these molecules of life and reconceptualized them as synthetic, i.e. as man-made, hybrid, and mobile chemical substances (for more on protein engineering, see chapter 4).

Mastering and playing with molecules

Many of the strategies to describe what happens to and within a protein during function, relied on interventions with molecular dynamics—insofar, my variation of Hacking's theme that if you can block a molecular pump, it must be real, also paid out an epistemic dividend—if you can block it, you can possibly understand it.[350]

One such strategy was "site-directed mutagenesis," i.e., the targeted and selective modification of a nucleotide triplet in a gene coding for one amino acid in order to exchange it for another. Site-directed mutagenesis allowed, for example, to swap a guanine-adenine-thymine in the DNA, coding for an aspartic acid in a protein, to an adenine-adenine-thymine, coding for an asparagine. Such specific mutations, long discussed in the context of designing biological function, had become possible since 1978, when synthesized DNA oligomers (the short snippets of DNA) were employed to change the respective codons in a gene—here is a direct connection to Khorana's work on nucleotide synthesis and its automatization in the gene machines of the early 1980s.

The rationale behind site-directed mutagenesis in the study of protein function was to exchange amino acids suspected as functional elements and to monitor the physiological effect; that is, to approach the

problem of what element did what in a protein by scrutinizing potentially changed or dysfunctional mutant proteins.[351] Within the mechanical framework of understanding membrane proteins emerging since
the early 1970s, the central question about the dynamics of the proton
pump was how this molecule used the energy of light-induced chemical
changes to transfer its freight, the proton, across the membrane, or, put
simply, what made it a pump. Spectroscopic analyses had revealed that a
certain type of amino acid, aspartic acids, which occurred several times
throughout bacteriorhodopsin's sequence existed in both a protonated
as well as a de-protonated state, i.e., these aspartates were able to bind
or release exactly the suspected freight. In a labor-intensive mutation
screen, Khorana used his convenience gene to construct 13 modified
variants (alleles) of the BR gene, in which the aspartate residues were
exchanged one by one for asparagines, a structurally similar amino acid
unable to perform the de- and reprotonation effect. The recombinant
proteins were expressed and biochemically purified from recombinant
E. coli host cells, and after chemical activation, these mutated pumps
were checked in an in vitro assay for function.[352] To make a long and sophisticated story short, this approach, in conjunction with studies from
other groups, suggested two specific aspartates at positions 85 and 96
in bacteriorhodopsin's amino acid chain, as crucial for accepting and
releasing the proton, or for doing the pump work. In the contemporary
molecular-mechanical explanations quoted in the Introduction, these aspartate residues became described functionally as BR's "proton accep-
tor" and "donor," respectively.[353]

It is important to stress that Khorana's recombinant DNA approach
to pinpoint functional elements of the pump molecule was only one way
to address the problem of mapping protein function on specific amino
acids. Another, more established way of approaching structure-function
relationships was practiced since the early days of molecular genetics,
e.g., in the famous 1940s experiments by George Beadle and Edward
Tatum on nutritional mutants of *Neurospora* that had correlated genes
with enzymes and their specific physiological functions.[354] This approach consisted in screening randomly mutated organisms in order to
find physiologically modified cells that would harbor loss-of-function
alleles. In the quest to find the functional elements of the pump, Oesterhelt's department at the MPI of Biochemistry built on this approach,
when they added chemical mutagens to growing cells and sequenced the
DNA of the respective loss-of-function alleles in order to spot sites of
crucial mutations.[355]

Even if both the directed and random mutation approach converged in identifying the same elements as crucial, Khorana's method using genetically engineered proteins had an advantage: The set of mutated alleles, available through the molecular genetic infrastructure of vectors, cloned alleles, etc., could be used for other experimental strategies to address protein function. For "molecular cross-linking," another interventionist strategy to probe structure and dynamics, mutated amino acids were modified chemically in order to block their molecular movements. To illustrate this approach, and the analogy to mechanics it induced, one may imagine two mobile parts of a macroscopic device being linked by a bar or stick, and thereby prevented from moving. The important point in Khorana's case was that the linkage of different spots within the protein allowed those parts of it that moved with respect to each other to be identified throughout the protein's functional cycle; i.e., this approach allowed a dynamic molecular topology of the molecule to be established.[356] That is, in contrast to crystallography's *static* images of molecules, cross-linking allowed insights into the *dynamics* of the molecular pump during functioning in vitro or even in the living cell. Here's the epistemic dividend of Hacking's theme: Molecular motions and interactions were probed and specified by blocking them.

The available site-directed mutants also allowed biophysicists to attach magnetic or fluorescent tags or labels to the protein, which could be detected by optical or magnetic resonance spectroscopies. Such labels took the role of "light beacons" attached to the molecule, permitting spectroscopists to peek into the molecular dynamics and surroundings at defined points of the molecule. Historically, this approach can be seen as an extension of the work pioneered by biochemist Mildred Cohn in Philadelphia in the 1960s, who had used metal ions in biological substances as probes for NMR spectroscopy.[357] Again in addition, and sometimes in opposition, to the static snapshots of protein structure from crystallography, these techniques helped to shape a dynamic picture of "what happened" within a biological molecule such as a pump, a receptor, or a transporter at a molecular resolution, and thus how structure was related to function.[358] Pinpointing episodes of what happened during the protein's functional cycle (after illumination, addition of substrates, etc.) to certain of its amino acids or even atoms, suggested a *functional topology* of the molecule, its *divisibility into parts* with discrete functions, at the same time as it enhanced the temporal resolution of its dynamics.

Overall, this precipitated in data on conformational change of a molecule when it performed its biological function. Knowing that the amino

acid at position 85, for example, was bacteriorhodopsin's proton donor fed into the pool of data that could be used to construct a sequential explanation of molecular action—this amino acid pushed this one,
transferred a proton from here to there, flipping around a domain, etc.
Studies such as the ones described here are where the Shakespearean narratives of molecular function described in the Introduction have resulted
from, the actors of which were atoms and chemical bonds, and which
became puzzling in their individuality and complexity as more and more
data were gathered.

Conclusion I: Plug-and-play, mechanisms, and the integration toward the molecular life sciences

Whereas BR had been a fortunate exception as a "user-friendly" natural
substance in the 1970s, attracting Racker, Khorana, and many others as
a material opportunity to devise new approaches and apply new technique, the situation changed after c. 1980: An increasingly large, multinational community of researchers had adopted recombinant DNA and
other chemical methods described above, turning this, but many other
proteins from scarce, delicate, and potentially impure natural substances
to such more akin to materials of chemical laboratories, which could be
produced or even synthesized under controlled conditions, modified, designed, and reassembled. Insofar, BR pioneered a trend that has become
ever broader since around 1980.

The set-up of a molecular genetic infrastructure, in combination with
biochemical and biophysical experiments such as cross-linking or spectroscopies, have allowed researchers to spell out molecular-mechanisms
of proteins in the ways that dominate today. This would have been impossible on the basis of proteins as natural substances, before these techniques took hold. In turn, this means that today's molecular-mechanical
perspective of life unfolded on the basis of a *transformed materiality of
proteins*: It was the joint use of synthetic methods from organic chemistry and recombinant DNA that allowed researchers to express protein
dynamics as molecular mechanisms and thus to turn them into what
they are conceived of today, molecular machines–on this point, Racker's
plug-and-play biology and Khorana's makings and probings of molecules are not only illustrative, but to a degree exemplary.

The research to spell out protein molecular mechanisms adumbrates a
novel way in which the molecular life sciences were practiced in the 1980s
and 1990s: The BR story as told in this and the preceding chapter has

shown that molecular genetics, structural biology, organic or biological chemistry, and biophysics remained in many respects disconnected fields in the 1970s, and that these fields became integrated on the level of actors, technique, and objects only after 1980—research on BR then encompassed biochemistry (protein purification, enzymology), membrane- or electrophysiology, structural biology (crystallography, EM), biophysics (fluorescence, Raman, electron and nuclear magnetic resonance spectroscopies), and recombinant DNA (cloning, mutagenesis).[359]

The increasing joint use of different physical, chemical, genetic, or physiological methods on one object, sometimes in a study of one group, sometimes by collaborating groups, meant not only that different researchers zoomed in on one object from different angles, but that they reciprocally built on each other's insights, in the sense that recombinant DNA allowed mutants to be constructed for genetic analyses as well as for spectroscopy, spectroscopic data were correlated to structural biology, structural biology was used to select and design further mutants, and so forth. The general picture of such an integrated use of methods blurred the boundaries between biochemistry, biophysics, and molecular genetics, and it appears characteristic of the contemporary molecular life sciences. Similar developments could be shown for other cases, such as neurophysiology's ion channels or transporters; however, the sequence in which approaches were applied to different objects, and the exact timing, differed.[360]

The combination of biochemistry, biophysics, and molecular genetics to study select proteins allowed "triangulations" relating data on molecular structure to those on dynamics and function. These coalesced into the visualized dynamic mechanical models of protein function shown in fig. 1 and plate 1. Obviously, not all these triangulations were successful and contradictions have remained between data from different approaches, sometimes casting doubt on one technique, sometimes on the entire project of molecular-mechanical analyses (see Conclusion). However, after 1990, research on an increasing number of larger, more complex proteins from animals, often with medical significance, has converged on a molecular-mechanical picture of function as described in complex, Shakespearean narratives.

While I have focused primarily on the impact of experimentation on the transformation of proteins, developments on what one may call their natural history, i.e., the collecting and classifying of molecular diversity, also need to be taken into account in order to draw a more complete historical picture of the molecular life sciences after 1980. These years saw not only an increase in protein structures being determined by X-ray crystallography, but also the expansion of large-scale databases, such as the Brookhaven Protein Data Bank (for comparisons of their 3D structures), as well as the creation of generic graphic models to display them and their functional domains.[361] On the basis of genetic or protein sequence comparisons, protein taxonomies were created, such as the family of the "seven transmembrane receptors" (also called G-protein coupled receptors) comprising not only rhodopsins, but also hormone receptors or cellular signal transducers.[362] Through shared sequence motives or functional domains, these families have suggested a common evolutionary history of proteins and general patterns to relate structure to function, which allowed characterizing common types of mechanisms among their idiosyncratic diversity.

One central and recurring theme in the entanglement of (re)making proteins, researching them in a plug-and-play mode, and understanding them mechanistically has been their *mastery* and *control*. Many researchers chose their objects of study as these lent themselves to experimentation in one or the other way, or as one could say, as they were "well-behaved." Some of the abovementioned studies also resemble playful tinkering with a molecule, as one may probe the mechanical interactions of an unknown contraption by unscrewing parts of it, bending them, fixing them, putting them together in a new order, et cetera. Mastering a biomolecule, having it as a pure substance in the test tube, and being able to use it for diverse experimental approaches was interpreted by the researchers as satisfaction, such as when Oesterhelt stated, in a conversation about his current research on a protein, "now that we master the fatty-acid synthetase [i.e., a protein complex] in all its aspects [. . .], it's huge fun."[363] Plug-and-play, or control in connection with play—monitoring color changes the test tube, recording traces of molecular action on papers, as seen in the previous chapter—reconnects to the project of synthesis as we have encountered it in Racker's or Khorana's cases: Making and mastering the cell's molecular inventory, piecing components together, and modifying them to understand them ultimately did not only shape the molecular-mechanical vision on an epistemic level, but these ways of acting on and interacting with proteins opened the door to thinking of them as molecular machines that could be actually put to use. On the level of organisms, human beings, or even societies, the objects and subjects of which were to be controlled through their molecular biology, the trope of mastery also harks back to the outset of what Lily Kay has called the "molecular of vision of life" (see Introduction).

Conclusion II: From making molecules and cells to synthetic biology?
A genealogy of practices in between chemistry and the life sciences

Another way of framing the research described in this chapter is to ask for its relevance for a very recent field, that of synthetic biology. Simplifying the developments and relationships of organic chemistry and the life sciences over the last four decades quite a bit, one could say that as the latter have become more chemical, the former has become more biological. The first part of this statement should be fairly clear: With the methods of recombinant DNA, that is, gene splicing and molecular cloning by tools such as restriction enzymes or synthetic probes, or the use of sequencing technologies and physical instruments, chemical thinking and working within biological research has significantly increased. It is now commonplace even in botany or zoology (think of DNA-based taxonomy and the equipment needed for it—DNA preparation kits, gel electrophoreses, or PCR machines).

Khorana's "way up" from synthesizing small to ever larger biomolecules such as a gene, and his subsequent adoption of recombinant DNA to make proteins in order to understand their mechanisms, exemplify the perhaps less obvious way in which organic chemistry has become more biological: We have seen the example of an organic chemist moving toward the biological realm, regarding not only the objects analyzed and synthesized, but also the tools and methods employed. The "going biomimetic" of chemistry, to pick up a term used in the 1960s, implied that the soft and delicate ways in which cells made substances were emulated in the test tube, at the expense of chemists' harsh (hot, explosive, and corrosive) reagents. This may represent a larger trend: Oesterhelt, who had approached biological chemistry very much from the angle of organic or natural product chemistry, adopted recombinant DNA in his lab in the 1980s, and later became involved in both genome analysis and biotechnologies.[364] On the level of lab equipment and industry, enzymes and kits for molecular biological routines have been marketed by biotech as well as by chemical companies, with many of the latter forming "life science" branches, encompassing biotechnology, biomedicine, and agriculture.[365]

Nevertheless, the share of chemistry, and the practices of synthesis that it has brought into the recent life sciences, has been curiously underrated in its extant historiography, which is dominated by the development of genetics-based biotechnology. From the 1940s to 1970s, or so one narrative goes, molecular genetics was dominated by its first, analytic phase, in which the basic molecules of life such as DNA were identified and central cellular processes were spelled out. In the early 1970s, the second, biotechnological phase began, focusing on the re-programming of organisms by recombinant DNA, or genetic engineering. The turning point between these two phases has been pinpointed to the first DNA cloning experiments, such as those by Stanley Cohen and Herbert Boyer to cut, paste, and transfer genetic elements between different microbes or similar strategies employed by Paul Berg for viruses.[366] Interestingly, the story of Khorana's synthetic gene, or Bruce Merrifield's synthetic enzyme and the machine to make it, fall into the same period. However, they decenter the historical picture of how the life sciences became invested in making from genetics, and reveal a broader trend that reconnects to synthesis as the quintessentially chemical project to remake the molecular inventory of life. Through the remaking of, e.g., vitamins and hormones, this synthetic project was in full swing long before molecular genetics entered the stage, it remained around and the two became entangled with each other.

However, neither the chemical nor the molecular genetic project of remaking as described in this chapter were primarily invested in the making of substances for use or the "creation of life" as a feat in itself, but rather in gaining understanding through remaking. As total synthesis was meant to confirm molecular structure to chemists, and as one synthesized element served as a stepping stone to make the next, larger one, so did the "wet tools" of the cell: Enzymes isolated in one round of investigation were used in further rounds of molecular genetic studies to obtain deeper insight into life on the molecular level. In instruments such as gene machines or DNA sequencers, the chemical and the biological making approaches have been integrated seamlessly, shaping the commercialized technological practice of the contemporary life sciences.[367]

So, what does this interplay and interlacing of chemical and biological making projects since c. 1970 tell us about the history of synthetic biology? This field, situated at the crossroads of, e.g., genomics, cell biology, and bioinformatics, has often been characterized by having imported a making ideal into the life sciences after 2000. Among synthetic biologists' portfolios range modest aims such as the production of substances by genetically and metabolically engineered organisms or the design of simple biologically functional elements such as a "switch," but also more ambitious ones as the construction of protocells, cell-like compartments delimited by a membrane, containing a hereditary system and a metabolism.[368] Whereas most protocells represent only short-lived aggregates to model biological processes, Craig Venter's institute actually plans to design artificial microbes, containing a minimal, engineered genome inserted into a membrane shell and cellular environment, which would serve biological energy production through modified enzymes. One ultimate goal of synthetic biology is the creation of novel forms of life, pieced together from modular components.[369]

It is striking to see that all of the material components required for a flagship project of contemporary synthetic biology such as Venter's protocells have been shaped in the research discussed in this chapter— Bangham and Racker's making of membranes and liposomes from isolated membrane lipids (the basis for a protocell), Khorana's synthesis of DNA and genetic elements by chemical methods, or the making and modification of proteins by recombinant DNA. That is, the chemical-molecular biological practices and the plug-and-play approach as described in this chapter can be read as a genealogy of practices to make and assemble components and parts of organisms leading up to synthetic biology. With reference to Michel Foucault's adaption of this Nietzschean concept, a genealogy does not depict a historical master narrative or origin (*Ursprung*) of recent synthetic biology, by following earlier emanations of the concept, and tracing their usage up to the present. By contrast, my focus on the scattered and small events and their rearrangement explain its contingent descent (*Herkunft*) from seemingly marginal fields and episodes of science.[370]

What could be gained historiographically by the genealogical perspective I am suggesting here? Existing historical accounts of synthetic biology have, on the one hand, highlighted its novelty as a coalescing discipline in the molecular life sciences of the turn of the millennium. On the other hand, the new synthetic biology has been pitched against prior usage of the term for attempts and announcements to "create new life from scratch," or at least to imitate its forms and functions. In many cases, these prior synthetic biologies have been pinpointed to novel insight into "life," and they frequently bordered on metaphysical questions and debates, such as in their first guise, the inorganic model systems of cells pioneered by French biologist Stéphane Leduc in the early twentieth century, centering more on morphology than on molecular composition and dynamics.[371] By contrast, a genealogy of practices feeding into synthetic biology is not primarily oriented at the proclaimed aim to "create life," but at ways of making biological molecules and compounds that have been shaped in between different fields of the chemical and the life sciences. This perspective brings into view the often piecemeal and mundane innovations on how to make biological Stoff in the laboratory. Rather than looking for one or several moments of foundation, the genealogical perspective highlights the events scattered over a longer period of time from whose rearrangements and confluence current syn-thetic biology has resulted: From the synthesis of nucleic acids, genes, and proteins, to the piecing together of liposomes, and (not discussed here in more detail) methods of cell fusion or cloning. Such a historical picture has the advantage of highlighting the impact of actors and prac-tices that frequently go unnoticed in narratives of foundations or origins, among them the contributions of organic, physical, or colloid chemistry in the last five decades—Foucault has characterized genealogy as "gray, meticulous and patiently documentary."[372] This does not mean to deny the influence of, e.g., genomics or computational biology in the present (and possibly the future), but it brings into view how life's materiality has already changed in small but significant ways under our eyes in re-cent decades.

When the German organic chemist turned recombinant DNA en-thusiast Ernst-Ludwig Winnacker announced a "synthetic biology" in the early 1980s, for example in the news piece from which the epi-graph of this chapter has been taken, his intention was certainly to con-struct a past for his own field that linked it to the scientific and eco-nomic successes of nineteenth-century synthetic chemistry. Yet, in spite of problematic historiography, he may have had a point here: The focus of the article was the introduction of gene machines à la Khorana and Merrifield to Germany, that is, an innovation to speed up the making of biological Stoff.[373] That DNA as a pure chemical substance spat out by these machines was biologically "absolutely dead," as the chemist oper-ating the machine told the journalist, was trite. Yet, more interestingly, thirty-five years in hindsight, we see that biological matter as a product of synthesis machines has, among many other remade components men-tioned in this chapter, subverted the boundary between the chemical and the biological bit by bit, thereby indeed changing what life is made of nowadays, and making certain pieced together lumps of matter perhaps a little less dead.

4 Biochip Fever: Life and Technology in the 1980s

Our machines are disturbingly lively, and we ourselves frighteningly inert.

Donna Haraway, 1985, p. 69.

Around 2007, an internet video of an ordinary laboratory rat, distributed as supplementary material to a publication on neuroengineering, created a stir. The furry creature sitting in an empty plastic box had been hard-wired by a fiber-optic cable attached to its skull, supplying light to its brain's motor cortex. The mouse was arbitrarily exploring its surroundings, sniffing corners, taking a few steps to the right or left, when an LED attached to the cable was switched on, creating a blueish halo around the animal's skull. Concurrently, the mouse began to run in anticlockwise circles rapidly and mechanically, but without panic or irregularity. When the light was switched off a few laps later, the animal resumed its prior behavior, sniffing, idling, exploring, as if the light-induced run had never taken place, or as if no traces were left in its memory. Was this murine cyborg a beginning of a new era of controlling life, of coupling it to digital technology, as the flood of articles on this or similar experiments in outlets such as *Wired* or *The New Yorker* has insinuated?[374]

"The Beam of Light That Flips a Switch That Turns on the Brain," thus reported for example the *New York Times* about the uncanny study, which has become an icon of the rapidly developing field of "optogenetics," a neologism combining optical stimulation and genetic engineering.[375] The animal, whose behavior was manipulated by engineering light-sensitive channel and pump proteins into membranes of specific cells from the brain's motor cortex and triggering these by illumination, is in fact an emblematic example not only for optogenetics, but generally for attempts to couple organisms to digital technology in, e.g., bioelectronics.[376] My point for setting the scene of this chapter by the optogenetic mouse, however, is neither this field per se, nor the immediate history of an emerging constellation such as this one (which, however, has direct links to the story told in the preceding chapters).[377]

For this history of membranes and molecular machines, the current technological use of proteins in optogenetics, which has unfolded in the first decade of the new millennium on the basis of methods from membrane research, recombinant DNA, cell biology, and neuroscience, provides a frame and a motivation to explore and conceive of the history of an earlier conjuncture between life, molecules, and technology: Whereas *Wired* enthused in 2015 about optogenetics' "fascinating little machines" that switch on and off brain activity, in the climate of burgeoning 1980s biotech and early nanotech, a comparable magazine such as *Omni* had mused about "biochips" and "biocomputing."[378] Coupling organisms to electronic technologies was one of the aims discussed under these labels (such as in prosthetics); however, the visions reached much further, up to chips to be built from proteins, and other biological molecules that were to accomplish a novel, radically different way of computing. Let me stress again that I claim neither that optogenetics stands in historical continuity to 1980s biocomputing, nor that its immediate goals were the same (biomedical technologies within an organism in the present case, organic or organismic computing devices in the dry world of microelectronics in the 1980s). If also under very different scientific, economic, and medial circumstances, however, both have attempted to tackle life's molecular machinery for computing technologies, as chips or interfaces, and both have zoomed in on structured and active biological matter.

The promise of "lively machines," Jan Müggenburg's term for bionic technology, now on a molecular scale, should in both cases be seen as instances of what Robert Bud has called the "enduring dream" of biotechnologies, that is, the attraction that technologies based on life have repeatedly posed to scientists, companies, and government in the past century.[379] This dream obviously encompassed the idea of untapped economic potentials, but, maybe less obvious from recent historiography, the idea of a better (i.e., more efficient, more refined, smarter) because lifelike technology, not only in medicine or agriculture. Biocomputers, for example, were presented as a radical alternative to silicon microelectronics, with some similarities in rhetoric as to how optogenetics' technologies are advertised to replace drugs and electrodes used to study the brain or treat disorders. And in both cases, these fields have been wrapped in comparable fever pitches of medial attention, with life's molecular machinery starring in colorful stories rather than in dry scientific publications.

With regard to the guiding question of this book, in what ways science and technology have come to think of and remake our cells and bodies as a "collection of protein machines," this chapter will add biotechnology to the picture, both as a style of doing research and as an actual engineering effort. To that purpose, I will analyze both the promises and the lab realities of biotechnologies in the 1980s, and characterize the specificities of concepts of molecular machines at the crossroads of recombinant DNA, biophysics, emerging nanotechnology, and computing/microelectronics. This will bring me back to the question of what about proteins and membranes has inspired biotechnologists and how they have in turn reshaped the materiality of life in order to create chips containing protein "switches."

But there are more reasons why the history of biocomputing needs to be told: First, it serves as an antidote to a historiography of 1980s biotechnology that takes into view largely the economically successful, biomedical aspects of the field, such as the production of recombinant proteins (insulin, interferon, etc.). Biocomputing shows that 1980s biotech was much more than this—in addition to and beyond venture capital and pharma, it was also the idea of alternative technologies modeled on life. Whereas this vision has not been realized in a way comparable to its biomedical side, both the general aims and the specific projects discussed in this chapter have remained around as part of tech discourse ever since—mostly under the label of nanotechnologies. Which brings me to another aim of this chapter: By sketching first, the general climate and the conceptual framework of biocomputing, and then zooming in on one case history of an attempt to actually build a biochip (from the purple membrane), I will also put more flesh on the bones of the history of nanotechnologies or "molecular electronics" (ME), as the field was called at the time, which has so far centered either on important figures (such as Eric Drexler, and his ideas of self-assembling molecular machinery) or broader research programs. This focus brings in actors and discourse that differ from those readers of this book have encountered in previous chapters—to simplify, one could say that biochip research took membranes and molecular machines from labs in California (such as Stoeckenius' at UCSF) into Californian labs around the world, and from *The Journal of Biological Chemistry* into sci-tech "fanzines" such as *Omni*. These were places and media of what historian of 1980s futurism and nanotechnology Patrick McCray has called "visioneering," that is, a specific form of mixing technological speculation with engineering, of making things and forming communities, often working and communicating beyond the confines of mainstream science and technology.[380] Such places, however, existed in their respective local fashions also beyond the US: By taking into view discussions and developments of bacteriorhodopsin-based chips in West Germany and getting a glimpse on what happened in the Soviet Union, I will contrast the approaches to bio- and nanotechnologies in very different research environments. This will provide a richer and more complex picture of the biotechnological history of membranes and molecular machines.

Alternative computing

The February 1984 edition of *highTechnology*, a monthly computing magazine combining business aspects with pieces on technological developments, confronted its readership with a hypothetical next frontier of microelectronics: the biochip (fig. 18).

The aims of such chips were to replace conventional silicon semiconductors with carbon-based biomolecules, thereby increasing miniaturization. Among the broad spectrum of attempts to use biological materials or biological design principles that were discussed in *highTechnology* was a proposal for a robot vision device by theoretical biophysicist Michael Conrad, a figure whose name and institution, Wayne State University, has been repeatedly mentioned when it came to biological inspiration for novel technologies in the 1980s.[381]

Conrad's goal was to build a hybrid biochemical-electronic device that would mimic the visual and cognitive system outside of an organism. To this end, he sketched a three-dimensional structure, to be realized as a sandwich-like "chip" of membranes or films stacked on top of each other. This was clearly modeled on the retina's anatomy: A first, "wet" photochemical layer would contain photosensitive chemicals (the reader of this book may think of rhodopsins or other photoreceptors, and in fact, Conrad discussed similar cellular signal-transducing systems). This sensitive layer would transform optical into a chemical signals, which further elements of the chip, layers of "protein switches," would then amplify.[382] Finally, the signal produced by this layer—one

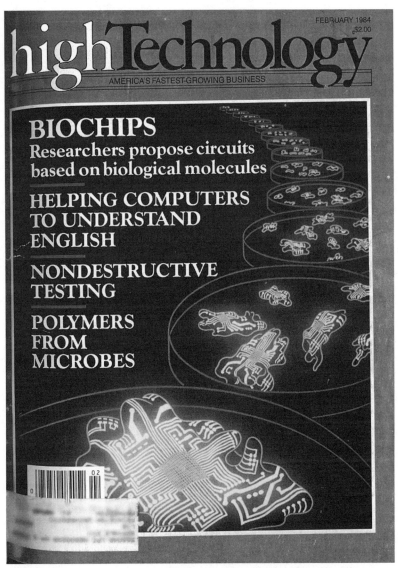

FEBRUARY 1984
$2.00

highTechnology
AMERICA'S FASTEST-GROWING BUSINESS

BIOCHIPS
Researchers propose circuits
based on biological molecules

HELPING COMPUTERS
TO UNDERSTAND
ENGLISH

NONDESTRUCTIVE
TESTING

POLYMERS
FROM
MICROBES

FIGURE 18 Merging microelectronics and biotech. Cover of *highTechnology* magazine, issue February 1984.

may think of a chemical reaction—would be transformed into an electrical one by an electrode covered with enzymes. Importantly, this chip's output would not comprise a "yes" or "no" answer as in digital systems, but a graded, approximate response, which was ultimately to inform a robot arm's reaction toward the perceived object. Apart from the final electrode component (a technology called "enzyme electrode"),

Conrad's biochip existed on paper only: It was a theoretical scheme informed by programming theory, physical chemistry, as well as biochemical and cell biological models of what should be possible in principle. As with many other proposals of the time, he remained vague on the issue of concrete substrates for his devices, mentioning different possible realizations.[383]

Conrad's robot vision device is a good example to introduce 1980s' biocomputing, not only due to its speculative nature, the envisaged makeup from biological components, or its reliance on knowledge from different disciplines. In fact, this device epitomized many of the expectations of what a form of electronic technology taking its inspiration from life was to accomplish: Conrad's biochip would, for example, optically detect an object not bit after bit. Put in computing language, it would not process information *serially*, but just like an eye, it would process all input in *parallel*; that is, it would detect structures altogether through their shape or gestalt. Such parallel processing, modeled also on how the brain was thought to work, was a much-discussed trope of 1980s hardware developments, as it promised to enhance computational capabilities. The approach was often explained by analogies to brain architecture.[384]

A second conceptual inspiration that Conrad's device took from life on the molecular scale was that it promised to mimic a way of cognition that was seen as typical of how biological systems processed information (to adopt the actors' language): As mentioned, biological molecules as computing devices would say not only "yes" or "one" and "no" or "zero" as did the silicon transistors on digital chips that embodied the binary of Boolean logics. Biochips, by contrast, were capable of saying "maybe," as explained in the *highTechnology* article, when, for example, a slight chemical reaction was induced by a stimulus to an enzyme.[385] This matter of degree was seen as an advantage over the rigid information processing of existing semiconductor technology, especially when it came to constructing optical interfaces or artificial intelligence.

In sum, the biochips envisaged in the 1980s could be tentatively characterized as computing devices made from biological materials and/or based on principles of biological design. They promised, among many other things, enhanced data density by providing connectivity not in two but in three dimensions or modes of computing that were entirely different from semiconductors, such as by molecular movements (conformational changes of proteins, see chapters 2 and 3) or optical effects (reading and writing by lasers).[386] Let me finish this introduction by saying what the biochips or biocomputers that will be subject of this chapter were not, since these and related terms have been used for very different things in the past three decades: 1980s biochips were not directly related to the so-called DNA computer, and they have even less in common with today's DNA chip technologies as used in biomedical diagnostics (although all these approaches tackle biological materials to address problems of detection or information processing).[387]

Beyond silicon — lifelike electronics

The *highTechnology* article sits amidst a fever pitch of popularizing accounts of biocomputing in the first half of the 1980s, which hyped the field as the coming merger of microelectronics and biotechnology.[388] In this way, biocomputing brought together two central motives and motivations of contemporary discourse on new technologies — first the promise of lifelike devices, and second the improvement of microelectronics by miniaturization — from switching units to chips to entire computers. Under the heading of "Moore's law," implying that the amount of circuit components that manufacturers were able to put onto a given area doubled within a period of a year or two, miniaturization has been widely discussed since the 1970s. According to historian of science Cyrus Mody, Moore's law should be understood not only from a technologically determinist perspective, that is, as a description of the progress in producing and arranging ever smaller switching units on microchips. The rule of thumb that came under this name equally comprised a performative dimension: As a social fact, Moore's law drove actors of research and development toward further miniaturization and innovation.[389] One field influenced by the permanently pending doom of ultimate miniaturization, was molecular electronics (ME). This loosely organized, iridescent field, comprising different novel approaches to microelectronics straddling science and engineering, gravitated around the general idea of using not macroscopic devices such as transistors, but molecules or even atoms as switches for computing. In other words, the basic units of microelectronics would not be made from bulk solid-state material, such as the layers of doped silicon in common transistors or integrated circuits. In an envisaged molecular computer, the device alternating between the two states forming a digital bit of information would be provided by a single molecule. Obviously, such a computer promised an extreme degree of miniaturization and therefore increases in capacity and power.

The history of ME as an alternative to silicon semiconductors stretches back to the 1950s, involving, e.g., physicist Arthur von Hippel and institutions such as the US Airforce or the Westinghouse corporation.[390] However, it was molecular electronics' second coming since the mid-1970s that brings us closer to understanding how molecules were thought to compute, and why biomolecules received specific attention. In the mid-1970s, researchers at IBM's labs in Yorktown Heights, New York City, and San Jose in California were studying what a central protagonist, physicist Arieh Aviram, called "molecular rectifiers." These were organic, i.e., carbon-based molecules that could function like a diode by conducting electricity in only one direction. As diodes were normally built from bulk semiconductors, and in a way formed the basis of a transistor, molecular rectifiers gave rise to a more generalized scheme of "organic electronics," that is, of synthesized carbon molecules serving as digital switches on an ultra small scale. In debates about the possibility of such molecular technology, Aviram justified their existence with evidence from biology—nature would have "developed comparable miniaturization several billion years ago in living organisms," whereby he referred to signal transmission in nerve (and thus, probably, synapses, ion channels, or the like).[391] Aviram's was certainly not a mainstream idea, but as we will see, there were a number of life scientists, some more senior and established, who formulated similar ideas around the time.

Some of the heterogeneous approaches falling under the heading of ME promised even more than miniaturization or switches beyond silicon: Their plan was to rethink central premises of computer architecture. On the conceptual level, computers in the second half of the twentieth century worked according to an architecture named after Hungarian-American mathematician John von Neumann. This referred to a device in which commands and data were represented as separate units stored on the same memory device. The basic operation cycle of a von Neumann computer would be to fetch a command and execute it on data picked up from the memory, before proceeding to the next operation. This sequence is the conceptual core of what has been described above as the serial mode of processing.[392] Attempts to move beyond the limits of serial processing loomed large around 1980, for instance, in projects aiming to process large amounts of data in parallel, as required for problems of artificial intelligence or automated cognition (such as a gestalt-like perception of figures, called pattern recognition). At the level of hardware development, parallel computing architectures aimed at implementing these goals were tackled in large-scale technological programs to further develop existing silicon-based microelectronics, such as the Japanese "Fifth generation project," and its American response, the "Strategic Computing Initiative."[393] Another approach to

solve the problem of parallel processing was the attempted use of macromolecules, i.e., not small synthetic carbon compounds as in Aviram's
case, but proteins as switches. Arrangements of such biomolecules were
thought to carry out parallel, non-binary (i.e., analogue) computing, as
in Conrad's abovementioned scheme. What is more, they subverted the
central distinction of a von Neumann computer, as the macromolecule
represented at once program (instruction materialized in its structure)
and data (information through its shape or conformation). With the previous chapters in mind, it is easy to imagine how enzymes and membrane
proteins as discrete molecular objects accomplishing their biological
function by structural rearrangements may have inspired such ideas—
here was a form of active or, metaphorically speaking, "programmed"
matter seemingly made to compute.

The second wave of ME, encompassing broad, diverse, and far-
reaching plans for a novel microelectronics, coalesced in the early 1980s
at a series of workshops called "Molecular Electronic Devices" (MED),
organized by Forrest Carter. The physicist, characterized by Mody as
a charismatic central figure of the emerging scene, held a PhD from
Caltech (where he reportedly partied with Richard Feynman, a central
post hoc inspiration for nanotech) and later worked on organic conducting materials at the US Naval Research Laboratories. Carter's ideas
inspired the MED gatherings, the stated purpose of which was, as he
expressed it in most general terms, "to explore the possibilities of developing switches at the molecular level for ultimately controlling and modifying signals."[394] Although many of the contributions to the first 1981
MED workshop centered on physical phenomena or synthetic molecules
(such as Aviram's), examples from the molecular life sciences, such as
photosynthesis, vision, or self-organizing cellular protein structures,
were also present. Biology became a more important strand of molecular
electronics throughout the second MED meeting: Here, Eric Drexler, a
charismatic MIT engineer from a younger generation than Carter, who
stood in for a cornucopian take on future technological developments,
presented a vision of "molecular machinery" that would earn him fame
under the label of nanotechnology in the years to come.[395] Like Aviram, Drexler took the existing molecular biological "technology," i.e.,
the protein and nucleic acids of the cell and their doings, as a "feasibility proof" for his far-reaching plans of miniaturization. Specifically,
his inspiration for the *Engines of Creation* (thus the title of his widely
distributed 1986 book) came from the prime examples of machine-like
macromolecules in the life sciences of the 1960s and 1970s as discussed
in the preceding chapters. His programmatic 1981 paper on "molecular

engineering" in the *Proceedings of the National Academy of Sciences* contains a table classifying "molecular machinery" (NB, this is an early example of the generic term) according to macroscopic devices—with the fibrous protein collagen representing a "cable," membrane proteins representing "pumps" to "move fluids," or the "genetic system" as a "numerical control system" to "store and read programs."[396] The specific point of fascination (and controversy) about Drexler's molecular machinery was that the existing small "tools" of the cell (such as the ribosome) were supposed to assemble even smaller, second-generation molecular machinery (later on, Drexler called this type of machines "assemblers"). The optimism that design of proteins and further miniaturization would become feasible was fueled not only by quoting Feynman's assessment that there was "plenty of room at the bottom," but on a material level by the recent advancements of recombinant DNA.[397]

Bio-inspired ME in the 1980s, that is, encompassed different actors than we have seen in this story previously, and unfolded in another scientific environment: Research on membrane structure, the function of enzymes or physical chemistry, which had taken place in rather conventional academic settings (carried out by people trained in the respective disciplines, presented at meetings of societies, or published in the respective specialized journals) suddenly appeared in the cosmos of guru-like personae such as Carter, or publicly visible, charismatic "visioneers" such as Drexler. Dry, expert matters such as biochemistry or biophysics, or even a technological approach to life such as recombinant DNA, turned into hotbeds for speculation of novel technological alternatives for microelectronics, but also medicine and other purposes. This was certainly catalyzed by the mix of actors gathering at MED (from physics, engineering, the life or computer sciences), but also by hip, technophile outlets in which this research was discussed: "The Biochip Revolution," for example, a piece penned by science writer Kathleen McAuliffe that appeared in the pages of *Omni* in 1981, may have actually coined this term.[398]

Omni was neither *Science*, nor the *Scientific American*: As a product of *Penthouse* publisher Robert Guccione and his wife, Kathy Keeton, the sleek and colorful sci-tech magazine shaped a characteristic 1980s' format for the popularization of science and technology, from (para)psychology to evolutionary biology, or recombinant DNA to microelectronics, and it became a characteristic outlet for visioneering à la Drexler.[399] McAuliffe's article, for example, introduced the biochip as a form of neuro-prosthetics, that is, an interface between the organic world and that of microelectronics, and quoted a start-up employee musing about the chip's development into "a compatible symbiote that will literally grow into the brain, establish communication with individual neurons, and thus learn from them in a biological sense."[400] For today's readers, this cyborg scenario may be more reminiscent of sci-fi writer William Gibson than of biotech history: Indeed, the second volume of the *Neuromancer* trilogy, *Count Zero* (1986), gravitates around the topic.[401] In addition to *Omni* and the daily press, biochips were discussed in other popular sciences magazines (such as the *New Scientist*), but as much as the topic was also a product of science-writing and a novel scientific persona such as the visioneer, there clearly was a more mundane dimension to it as well: Articles appeared in established scientific journals (albeit often without terms such as "biochip"), and there were outlets subverting the idea of clearly separated camps.[402]

The insufficiency of considering these development in a dichotomous framework of a countercultural, "groovy" and/or entrepreneurial versus a more conventional academe pertains not only to the media, but also to the actors: In spite of the importance of individuals such as Carter or Drexler, the MED meetings and molecular electronics generally were not solely an affair of gurus and their followers, and neither were they an American affair only, as the remainder of this chapter will show.[403] Contributions were authored by international scientists, such as the Swiss-German surface chemist Hans Kuhn, biochemist Hartmut Michel, or X-ray crystallographer Johann Deisenhofer. Whereas Kuhn was a director at the MPI of Biophysical Chemistry at Göttingen, later Nobel laureates Deisenhofer and Michel worked on membrane protein structure at the MPI, in the ambit of Oesterhelt. These were rather down-to-earth academics that did not smack of either counterculture or entrepreneurship. Moreover, projects similar to those present at MED were initiated in, e.g., West Germany or Japan.[404]

The questions of how the boundaries between the established fields of science and novel, cross-disciplinary activities such as ME developed, how reciprocal influences played out and how different individuals navigated this new situation, or how biocomputing projects developed in different countries will be followed up through the remainder of this chapter. But first let us see what were the inspirations, or the scientific points of departure, for schemes such as Conrad's biodevice or *Omni*'s biochips.

Membranes and proteins as biological technologies

Musing about biochips as future "brain symbiotes," or about cells being filled with a fleet of molecular machinery just waiting to be improved by humans, certainly did not represent the mainstream of how the molecular life sciences considered the potentials of protein technologies at the time. And yet, similar sounding ideas had been uttered a few years previously from more established quarters of science, and it may well be that these inspired visioneers such as Drexler.

Notably, the call for engineers to tackle the molecular biological machinery for microelectronics originated from membrane and protein science. "Living cells contain the ultimate in microelectronics," thus announced biophysicist Britton Chance in *Physics Today* of 1980, before detailing membranes and photosynthesis, vision or the function of mitochondria.[405] Chance, a well-established representative of postwar basic biophysics, based at the Johnson Foundation in Philadelphia, had been a central figure in bioenergetics. Against the background of his prior work, e.g., spectroscopic insight into how membrane proteins orchestrated molecular movements and electronic processes of metabolism (remember that he also adopted BR as a model for spectroscopic work in the 1970s), this article took a surprisingly visionary glance, that indeed foreshadowed tech discourse of the coming decade, for example, when he finished on the note that "assembled into a brain, for example, they [i.e., these biomolecules, M. G.] perform as incredibly powerful information processors. Physicists and engineers concerned with building ever smaller devices can look to them for inspiration."[406]

Such a bold statement from a senior basic researcher could be read in many ways. First of all, the journal needs to be taken into account—somewhat similar to *Scientific American*, *Physics Today* is neither a peer-reviewed journal of a discipline, nor does it speak to the broader public. It was a monthly communication of a scientific society that included generalizing pieces directed at physicists more broadly construed, as much as, possibly, to policy makers, science journalists, etc. Thus, a visionary article by a reputed individual such as Chance may have also had journalistic background inspired by, e.g., an editor. Taking the article's content literally, what it accomplished was to explain the new terrain of biophysics, membranes and proteins to physicists and engineers. Moreover, the article set out as a reply to an issue of the journal on "Microscience" one year previously. Whereas the 1979 issue, opening with a synopsis of Richard Feynman's famous lecture on microscopic technology in physics from twenty years before, Chance's was a biological reply, echoing what we have already heard from, e.g., Delbrück or Aviram, that "biological systems, however, have, in a sense, solved the problems associated with such small microstructures."[407]

Interestingly, Chance's paper is another early case where the generic term of "molecular machines" was used (and it was quoted by Drexler a year later). But as most of membrane research was not in any sense "applied" at the time, tackling quite basic issues, was Chance only speculating when he talked about technologies?[408]

In fact, a number of surface or membrane technologies that could be further developed for computing processes—think back to Conrad's sandwich-like biochip—existed, and someone tinkering in this area was Hans Kuhn, the Swiss chemist and MPI director who contributed to the second MED meeting. After completing post-docs with Niels Bohr and Linus Pauling, Kuhn became head of the department for "Molecular Systems Assembly" (*Molekularer Systemaufbau*) at the MPI for Biophysical Chemistry—the institute that was at the time home to self-organization theorist Manfred Eigen or neurobiologist and later patch-clamp Nobel prize winners Erwin Neher and Bert Sakmann. Kuhn used synthetic membrane models, so-called Langmuir-Blodgett (LB) films, roughly speaking a planar version of Bangham's liposomes, to assemble and study structured molecular layers, which became a frequently discussed building block for biochips among the MED community.[409] Kuhn's films comprised sandwich-like protein-lipid arrangements, which he had considered already in the 1960s "construction kits" to create "molecular switching devices" for "information processing."[410]

Chance and Kuhn illustrate how senior, mainstream scientists from biophysics or physical chemistry have supplied not only keywords, but concepts and material models to the forming biocomputing scene. Their work also illustrates how permeable the boundaries must have been around 1980 between established institutions and a realm of scientific-technological speculation that later gained a "louche reputation" among a more conventional academe.[411] Finally, these papers illustrate an unexpected point of departure for biocomputing, or ideas of molecular technologies in general: research on membranes, proteins, and their conformational changes, as well as the electrical and chemical dynamics of metabolism, seemed to suggest that a world of ordered molecular entities and processes existed in cells, discovered in the 1960s and 1970s, ready to be tackled technologically in the 1980s. Interestingly, biological objects were addressed here at their lowest level, i.e., that of atoms and chemical bonds in macromolecules studied by physical and chemical approaches. For a historian or a philosopher, this raises the question of what (if anything) was perceived as specifically biological about these molecules and their doings as compared to complex, man-made chemical substances, such as Aviram's organic rectifiers. However, let us first look at the "tools" envisaged to tackle this cosmos of biological molecules.

Cloning a computer—the ultimate scenario of recombinant DNA

Recombinant DNA around 1980 was a set of methods from molecular genetics by which it had become possibly to mobilize, modify, and produce proteins by acting on their genes with "wet" tools isolated from cells. To understand how certain researchers reacted to the novel procedures to splice and link genes (molecular cloning), to shuttle proteins between different organisms, or to introduce mutations into them (as discussed for Khorana's work in the preceding chapter), however, one has to also take into account the enthusiasm and the expectations (economical and others) that accompanied what became the first wave of molecular biotechnology. The few years when scientists turned "gene jockeys," to quote Nicolas Rasmussen, were fishing and cloning the DNA of hormones such as insulin and produced the substance artificially in bacteria introduced a new speed of research into the life sciences, driven by investment and patenting. These developments gave molecular biological research enhanced public visibility and political importance, and they helped shape a novel, entrepreneurial scientific persona.[412]

At the MED meetings, molecular geneticist Kevin Ulmer, representing a biotech start-up company by the name "Genex," formulated the expectations of what recombinant DNA could bring for biocomputing. As Aviram or Drexler before him, Ulmer took existing life as the proof of the principle that molecular technology was possible, and that the tools of genetic engineering would provide "the ability to biologically produce *any* protein we desire."[413] In order to reinforce this argument, he mentioned the design of functional enzymes from scratch, the prediction of protein 3-D structures, and other topics or problems of biochemistry and biophysics, which made it appear as if molecular technology seemed right around the corner (many of these issues have remained unresolved).

It is not easy to provide an adequate historical understanding of these generous expectations. In addition to the marketing aspect that was certainly inherent in Ulmer's statement, one has to take into account that in the early 1980s, protein engineering on the basis of recombinant DNA was indeed a rapidly developing field: Techniques were used that had been inconceivable a mere decade previously (DNA and protein synthesis), and bold prognoses extrapolating these past developments may have looked a lot more plausible than they do in retrospect.

Taking up the concepts of the German historian Reinhart Koselleck to understand the modern transformation of historical experience and expectation in the wake of accelerating social and technological developments, one could say that the "space of experience" (*Erfahrungsraum*) provided by recombinant DNA, that is, the fresh and forceful historical experience that objects and processes of life had become open to change, set in motion a broadening of the actor's "horizon of expectation" (*Erwartungshorizont*): The actual as much as the experienced tempo of research increased, and the "ultimate" was perceived as possible, if not imminent (this term was used by Chance, Ulmer, and other quite heterogeneous researchers).[414]

Yet, the nitty-gritty of genetic and protein engineering at the time reveals the flipside of this experience of acceleration: My account of Khorana's 1980s' work to clone and modify proteins has shown that recombinant DNA at the time comprised tedious, year-long work by specialists in cutting-edge institutions of science in order to clone single genes, to introduce a few specific mutations, or to have cells synthesize minute amounts of a recombinant protein. Often enough, such accomplishments represented proof-of-principle or demonstrative work rather than technological solutions to actually make proteins for use, let alone to change the world with them (it remained, for example, still easier to isolate a relatively well-studied model membrane protein such as rhodopsin from eyes obtained at a slaughterhouse than to produce it biotechnologically). In a similar vein, Nicolas Rasmussen has argued that with regard to commercial biomedical projects, the early revolutionary expectations were not met, and that the field underwent a normalization toward much longer and complicated research after the first "low-hanging fruit" of, e.g., cloned hormones had been reaped.[415]

In comparison with Khorana, Ulmer's "ultimate scenario"—to develop a "genome" for a computer that would "code" for some sort of protein-based circuitry self-assembling in a cell—clearly adumbrated a totally different level of feasibility. As discussions of his talk at the MED meetings suggest, some participants perceived such utterings as improper speculation at the time and were highly critical regarding both general conceptual problems and practical feasibility. Caltech biophysicist Howard Berg, for example, took an ironical stance on the issue, when he reportedly quipped that a computer based on biomaterials would have to be kept sterile in order to prevent microorganisms from "eating it."[416]

In retrospect, recombinant DNA certainly had given a strong stimulus to conceiving of biochips, as its routines to manipulate genes and proteins had opened up novel scenarios regarding what had *in principle* become possible to accomplish and what was expected to become possible. However, regarding concrete model systems or laboratory approaches to actually build biochips, or simple components of such devices, recombinant DNA had comparably less to offer until the mid- to late 1980s— actual experiments relied either on synthetic models that merely emulated certain properties of biological structures, such as Kuhn's LB-films, or simple existing models such as the purple membrane (see below). Most applications of recombinant DNA in, e.g., mutagenesis and protein engineering achieved only baby steps when compared to Ulmer's or McAuliffe's "ultimate scenarios."

In other words, recombinant DNA opened up a space of possibilities for biocomputing through the advent of different actors on the personal and institutional level (such as Ulmer or start-ups like Genex), by adding novel dimensions to existing projects in biochemistry and biophysics that did not aim at technology before, and, maybe most importantly, by providing an extended horizon of expectation of future developments— the ultimate appeared close enough to be taken into view. Thus, in addition to being merely one method among others or a way to make money, recombinant DNA should be understood equally as a ferment of research, accelerating it and driving it in novel directions. If membranes and proteins were the machines of the future, then recombinant DNA was the ferment to bring about that future. However, what were the envisaged gains of novel lifelike technologies?

Molecular bionics: Self-organization, evolution, and adaptation

In 1984, Kathleen McAuliffe penned another piece on biocomputing for *Omni*. "Smart cells" presented the "chief champion of a radical new view" on the topic: Stuart Hameroff, originally an anesthesiologist from Arizona, participant of the first MED workshop, was introduced as radical because he advocated the use of self-assembling cell structures called microtubuli as a basis for a mechanical, materialized form of computing.[417] Ultrastructure analyses of cells, such as by EM, had revealed microtubuli to be involved in the building of the cytoskeleton, but the fibrous protein assemblies were not only known as the cell's "microscopic bones"—they also served to organize intracellular transport and communication, forming a dynamic network of active structures (one microtubulus being formed from different protein subunits, which offered a large number of arrangements).

An animated model of one microtubulus as a cylindrical structure pieced together from small components seems to have even been on display

FIGURE 19 A radical new view of computing. Model of a microtubule presented at the first "Molecular Electronic Devices" meeting in 1981. The model was composed of black and white parts (protein subunits), and literal protein sidearms displaying molecular transfer processes in a bucket-brigade-like way. Technology and later quantum consciousness pioneer Stuart Hameroff conceived of such cellular processes as a basis for molecular-mechanical computing elements inspired by life. From: Carter, Forrest, ed. 1982. *Molecular Electronic Devices*. New York: Marcel Dekker. Figure 4. p. vi. Reproduced with permission.

at MED. Hameroff assumed that these structures were able to save and process information by rearrangements of the protein's subunits, which could exist in different "conformational" states (white and black in the model), similar to so-called cellular automata, a model from information theory much discussed in the context of "artificial life" at the time.[418] An entirely black cluster of subunits would thus represent one informational state of the structure, different combinations of white and black ones others, and so on (fig. 19).

Though certainly quirky, Hameroff was not the only one pondering whether microtubular self-assembly and reorganization, especially in nerve cells, represented a material basis for cognitive processes such as memory formation. *Omni* explained this analogy by the "uncanny resemblance" Hameroff had observed between their ultrastructure and that of a bubble memory, a computing element considered for a while as a competitor to conventional data storage technologies.[419] Extrapolating broadly from this morphological similarity, it was only a small leap to speak of microtubuli as "processors" within cells, in which information storage was accomplished by said molecular-mechanical reconfigurations of their subunits.

Hameroff's 1987 book, *Ultimate Computing: Biomolecular Consciousness and NanoTechnology*, described this theory in the format of a monograph, and in fact, a somewhat unconventional one: On the one side, the title's wording and orthography as much as the cover design (decorated with graphic patterns from microtubular elements) were more reminiscent of popularizing literature, such as *Omni*, than of a sober science title published with Elsevier (which it was).[420] On the other, this was clearly a monograph addressed to professionals in biophysics, computing theory, and engineering, demanding a high level of cross-disciplinary expertise, not least in mathematics, and offering comprehensive bibliographies of current research literature.

Ultimate Computing, reiterating the trope of imminent scientific and technological breakthrough, devised a vast, speculative picture of scientific and technological development: Based on the idea that the cytoskeleton (assembled from microtubuli) was "the cell's nervous system, *the* biological controller/computer," Hameroff rushed from the biochemistry of microtubular self-assembly to computing theory to consciousness, and finally to the evolution of a type of novel type of technology that was to mediate life and conventional human artifacts — "perhaps a merger of mind and machine: *Ultimate Computing*."[421] Whoever may dismiss Hameroff as merely an idiosyncratic voice should take into account that microtubular theories of cognition have remained around ever since, for example, as part of the 1990s debates about quantum consciousness, in which Hameroff got involved alongside British mathematician Roger Penrose.[422] Among those championing similar speculative scenarios to understand cognition as an effect of a dynamic, self-organizing, and possibly evolving material structure ranged also cell and microbiologist Lynn Margulis. Of Gaia fame for the 1970s work on the self-regulating planet with British physicist James Lovelock, Margulis was another characteristic and controversial figure of 1980s science: As much as the theory of the homeostatic earth had become a political argument and a widely known countercultural slogan, Margulis' "endosymbiotic theory" of cell evolution had moved from a renegade to a mainstream position, foreshadowing to some a radically different view on life as well. By postulating that all cells of higher organisms have resulted from cooperating smaller units (originally symbiotic bacteria engulfed by a larger cell), endosymbiosis challenged the fundaments of a neo-Darwinian evolution "red in tooth and claw."[423]

Turning symbiosis into a general principle of life, Margulis took the microtubular theory one step further: She argued that just like the cellular nucleus, microtubuli were evolutionary remnants of microbial symbionts taken up by cells a long time ago, and that these still discernible and somewhat autonomous units accomplished cognition (from perception in microbes to the human mind).[424] In her model of a "symbiotic brain," cognition was thus achieved through these structures' molecular dynamics, similar to Hameroff's model. Put forth in her 1986 popular monograph, *Microcosmos: 4 Billion Years of Evolution from Our Microbial Ancestors* (coauthored with her son Dorian Sagan), the theory of the symbiotic brain was received predominantly critically among peers, which, however, did not prevent the authors from concocting truly mind-boggling bio-technological scenarios. These put Margulis as a feminist, countercultural life scientist in the somewhat unlikely company of Hameroff, or technological futurists such as Drexler or Ulmer:

> As computers and machines come together in the new field of robotics, so robotics and bacteria may ultimately be united in the so-called "biochip," based not on silicon but on complex organic compounds, that is, an organic computer. Like plants performing photosynthesis, these manufactured molecules would exchange energy with their surroundings. But rather than turning it into cell material, they would turn it into information. [. . .] The outcome of information exchange between computer, robotic and biological technologies is not foreseeable. Perhaps only the most outlandish predictions have any chance of coming true.[425]

The general concept that informed such hypotheses about mergers of life and machines under the umbrella of a novel form of technology could be characterized as a molecular version of bionics: Here, the role of the biological was neither only to provide new materials to solve problems such as miniaturization, nor simply the expansion of technological feasibility by recombinant DNA. Molecular bionics was about taking inspiration for technologies from life by adopting principles of biological function at the level of cells or molecules, and thereby subverting the boundary between them.[426]

Macrolevel bionics had existed throughout the twentieth century, e.g., in designs of artifacts following form and function of biological objects such as wings and the strategy to model technologies on organisms was taken up in postwar cybernetic perception studies, such as the design of an artificial retina. However, a molecular version of bionics seems specific to the 1980s: Attempts to transfer specific biological principles of function to the realm of human technology had moved to a novel level.[427] Thus, it is no coincidence that Hameroff quoted Buckminster Fuller's geodesic domes or holography as paragons for what was to be achieved with microtubules. Interacting networks of cell-like structures or of biomolecules, again in conjunction with the extended horizon of expectation provided by recombinant DNA, foreshadowed a kind of technology that would supposedly differ on a qualitative level from the existing man-made artifacts and that was intended to bridge the perceived gap between the living and the inanimate. This was to lead to a bio-technological "co-evolution" of organisms and technology— tellingly this biological concept was also the title of a journal published by Stewart Brand since the mid-1970s, and it became a slogan for a new way of life and alternatives of social development.[428]

The complexity and adaptability of life were other motives that biocomputing advocates took up in this context: It was because biological systems were capable of going beyond Boolean logics, or rigid programmability, as described above, that they would become adaptable, which in turn opened up the possibility of technological evolution. Michael Conrad, for example, argued that enzymes or antibodies detecting a substrate by a lock-and-key type mechanism could be gradually modified by recombinant DNA in order to become adaptable.[429] And the quoted *highTechnology* article thus mused about a "natural relationship" between the material structure of such analogue computers and their tasks that was superior at complex, context-dependent data processing.[430]

In spite of all differences regarding scientific background as well as the concreteness and articulation of their schemes, a commonality between the microtubules of Hameroff and Margulis versus Michael Conrad's biochips was that these approaches conceived of computing as a *material process*: Signals did not remain immaterial electric processes any longer, since a dynamic and specifically shaped physical structure would process information in computing proteins. The distinction between data and program would become obsolete.

In other words, the materiality of computing and information, and as will be shown below, specifically active protein and membrane matter, loomed large in what biology seemed to offer to microelectronics. For Conrad, such complex, sensitive material structures were to become "augmentations" of existing technology toward a more lifelike behavior. Historically, or so he concluded, the age of biocomputers as a return to life's technology would conceive of their programmable antecedents as an "artificial add-on of the twentieth century."[431] Yet, the molecular technology aimed at turned out to be problematic in many respects. As Michael Conrad correctly observed, the complexity required for technological systems to be open to adaptation or evolution was also what made them difficult to control or manipulate, leading to a dearth of models to begin with.

The visionary, alternative, and radical aspects of 1980s biotech discussed in this chapter, however outlandish they may appear today, stand in stark contrast to the received winner's narrative of venture capital and biomedicine. However, these actors, their ideas, and their medial representations were also part of the inextricable network of plotting, planning, making, and projecting, in short of biotech discourse and tinkering at the time. Membrane and protein science as a substrate of biocomputing, recombinant DNA as its ferment, and molecular bionics trying to mimic hallmarks of life such as self-organization, adaptability, and evolution were "bio-tech" in a more radical sense, attempting to bridge a perceived gap between existing physical or chemical technology and life. Similar themes surfaced not only in biocomputing, but also in ideas for biological energy production, or when it came to the environmental aspect of technologies.[432]

As Robert Bud's study of the various encounters between life and technology throughout the twentieth century has shown, this broader meaning of biotech was all but new beyond the molecular level. Yet, detecting it also in the 1980s, among biomedicine and the entrepreneurial scientist, unexpectedly connected the protagonists of this chapter with discourses about alternative technologies reaching back to the interwar period.[433] The fact that the vast majority of ideas or model systems of biocomputing, such as Kuhn's LB-films, have never been turned into any type of marketable product should not make us forget that, to obtain a richer historical picture of what biotechnology represented and promised around 1980, these stories must be added to a historiography that focuses all too often on projects with immediate economic or technological fallout.

From protein to prototype: Materializing a "molecular switch"

The 1984 *highTechnology* piece on computing molecules finished on a sobering note. The response from the American microelectronics industry to schemes such as Michael Conrad's had been "nil," and to secure funding, the "basic feasibility of the technology" had to be shown.[434] Even a biochip enthusiast such as Felix T. Hong, who presented with Conrad at the third MED meeting (1988) and edited a volume on "Biosensors and Biocomputers," had to concede that experiments lagged far behind theoreticians' "hot pursuit." What were needed were data and prototypes, serving as "feeders" to computer scientists. Arguably the most important model system for a molecular computing element in the 1980s was BR, and the reasons listed by Hong sound familiar: He took up the concept shaped in the 1970s of the protein as a "pump" or "photosynthetic device" that could serve as a "prototype for biochip research and development," whereas the related visual system represented a "light-activated switch, similar to a phototransistor."[435] Of interest where the abilities of these molecules to react to light by changing conformation (switching) or by producing a photoelectrical effect (pumping). These features represented "by nature" what those interested in biological computers were looking for. Moreover, BR seemed to lend itself to technological uses: The substance was easily prepared and purified; it remained active for a year in an artificial environment; it was one of the best understood systems; and it displayed sufficient complexity to be engineered by the methods of protein chemistry or recombinant DNA.

Hong was not alone in this assessment: In the US, but also in West Germany, Hungary, and the Soviet Union, scientists-as-visioneers, conventional academe, and industry attempted to engineer this molecular pump or switch.[436] In analyzing the development of the German project based at Munich I pursue several goals. First, my account adds a concrete case from the laboratory bench to the histories of 1980s' molecular technologies. Even as this case story directly connects to biocomputing as introduced above, it involved very different actors and another environment of technological development. Contrasting this attempted materialization of a "biomolecular switch" with the field's programmatic will reveal differing perceptions of its economic and technological potentials as well as problems and contradictions faced by those who attempted to turn molecules into technologies.

Bearing in mind the fancy articles in *Omni*, or the colorful blend of scientists and tech-geeks who gathered at the MED meetings, the project that probably took the idea of computing biological molecules furthest toward realization in the 1980s appears rather down to earth, if not, frankly, a little boring. When the company newsletter of Wacker Chemie, a Bavarian chemical company specializing in the production of bulk materials from chlorine to silicon, announced this new project in 1989, the heading somewhat dryly announced a "biopolymer for optical information processing" rather than a biochip.[437] In spite of this different language, which addressed industrial chemists rather than visioneers or the public, the goal of an established company listed on the stock exchange was quite similar to what Conrad's sandwich-like chip or Hong's assays would have looked like: A joint team of researchers from Munich University, the MPI for Biochemistry (where Oesterhelt's department specialized in membrane proteins), and Wacker's in-house research department, the "Consortium für elektrochemische Industrie," had worked since the mid-1980s on an optical information processing device, the centerpiece of which represented an analogue "chip" based on BR. In fact, this computing unit would more technically be called a film, as it consisted of oriented pump molecules within a polymer layer, similar to biological membranes or, even closer, to the LB-films that Hans Kuhn had presented at the MED meetings. Based on the molecule's characteristic interchange between two photostates, the aim was to use this film as a medium for reversible information storage: A green laser (absorbed by the purple state of the molecule, "switching" it to yellow) would thus write the information into the film, with, in principle, every molecule figuring as one bit, while a red laser would read out the information. As the pictures in the article illustrate, the instrument did not look "biological" at all—the "biochip" was a semi-transparent purple foil, and the rest of the device consisted of a coherent optics laser set-up, which required lenses, mirrors, and light sources to be installed with high precision on a solid optical table.

The Munich biochip had several intended purposes: In addition to its use for reversible, high density optical data storage, it should also be applied as a processor of optical information—for automated vision, as in Conrad's envisaged device. Wacker's cooperation with physical chemists and biochemists aimed at constructing an instrument that would compare video images in real-time and thereby accomplish pattern recognition in images—among the mentioned purposes of such a device ranged meteorology (dynamic analysis of cloud images), non-destructive materials testing (e.g., detection of fissures in a work piece), or, as a military application, automated so-called friend-foe detection. The innovation of the "real-time holographic image correlator," which existed as a laboratory set-up, was that it did not transform images bit by bit into digital signals, but recorded, read, and compared them in a gestalt-like way all at once. In other words, this device materialized a form of the much-discussed parallel computing—built, however, not by tech prophets or a start-up, but by an existing chemical company's research facility and conventional academics.

Biotech and molecular electronics in West Germany

It may come as a surprise that actors who differ in many respects from the colorful folks met in the first half of this chapter zoomed in on a very similar project. To understand the background of Wacker's foray into biocomputing under another name a little better, it is necessary to trace the early history of biotech in Germany.

Whereas American biotech is associated with venture capital and the formation of start-up companies such as Genentech, or more closely related to biochips, the short-lived Genex of Kevin Ulmer, West Germany took a different path. Reasons were not only the differences in science policy and university structure (all of which were public sector institutions), but the dominance of established chemical companies (the big three being BASF, Bayer, and Hoechst) and their strategies to make substances from bulk to fine chemicals by petrochemical syntheses. Moreover, the country was clearly lagging behind the US when it came to recombinant DNA research around 1980 — one pivotal moment of biotech in Germany thus was the decision of the Frankfurt-based pharmaceutical company Hoechst to fund a genetic engineering facility at Boston's Massachusetts General Hospital rather than at home.[438]

The impression arose that Germany was missing out on an important novel technology that was thought to counteract the loss of classical industrial production, which had affected the country as much as other Western European ones or the US. In consequence, recombinant DNA became a factor of technology politics: It promised to create new jobs, to decrease dependency on oil imports required for petrochemical syntheses, and it was considered at least by some as a more environmental way of production.

Under these premises, the federal government remodeled a major state-funded research and development (R&D) program for biotechnology that had already existed through the 1970s. In the first phase of the program, funds were handed out to big companies or state research facilities for a variety of projects, many of them directly applied, such as in food or energy production. Under the impression of recombinant DNA and Hoechst's decision, more basic research was included, and grants to create partnerships between research establishments and companies were handed out (the start-up model existed as well, but to a lesser extent). Moreover, this scheme involved the creation of four *"Genzentren"* (gene centers), new institutions to facilitate technology transfer between academia and industry in recombinant DNA matters. In Munich, the Genzentrum was housed in the MPI of Biochemistry, its first director being the organic chemist turned 1980s' "synthetic biologist" Ernst-Ludwig Winnacker.[439] In a climate of reinforced state funding for projects of molecular biotechnology, including ones with a high risk (the portfolio of the federal program ranged from biomedicine to novel areas such as "bioelectronics"), Dieter Oesterhelt, director at the MPI of Biochemistry since 1979, garnered support for the project and the head of Wacker's research department. Together with physical chemist Norbert Hampp from Munich University, a joint "Verbundprojekt Biosensorik" (collaborative research project on biosensorics) was launched, receiving c. 2.7 million Deutsche Mark in funding from 1985 to 1991.[440]

It may seem strange to see a producer of materials such as Wacker Chemie involved in a biocomputing project. However, Wacker's contribution, which consisted chiefly in devising methods to produce biological materials, fit into its larger transformation, with the company beginning to adopt biotechnology at this time. Several cooperative efforts between the company, the Genzentrum, and the university were formed, all centering on the production of complex organic substances, so-called fine chemicals (as opposed to bulk substances), by fermentation culture, a classic biotechnological approach since the days of penicillin production. In the recombinant DNA era, the use of genetically engineered bacteria for fermentation suggested that it would become possible to manufacture all kinds of substances not by synthetic chemistry, but using engineered physiological processes of organisms. Moreover, as pure silicon for semiconductors had ranged amongst Wacker's important products, it makes sense that the company developed an interest in materials that promised to be an alternative for information technologies. To simplify slightly, a company known for chlorine, plastics, silicon, and rocket fuel took up food additives, pharmaceuticals, and films for information processing, with the business reports of the early 1990s proudly presenting fermentation vats, Petri dishes, and brownish bacterial cultures instead of shiny synthetics.[441]

Comparable reorientations toward biotechnological products and ways of making in the 1980s took place in many other companies as well—in pharma, Schering and Bayer, for example, expanded their portfolios toward "red" pharmaceutical biotech, with the latter rebranding itself as a "life science company" later. Producers of materials comparable to Wacker, such as the Belgian Solvay company, adopted "white" industrial biotech, and a company that is now synonymous for "green" plant biotech, Monsanto, had previously been a producer of silicon for microelectronics, among many other substances and materials.[442]

The biochip project was subdivided so that the company's new, experimental biotechnological facility developed production methods for substances, while the academic partners focused on screening for mutants of the protein suitable for optical uses (Oesterhelt's department at the MPI) or the set-up of the laser optical device (Hampp at Munich University).

Publications in physics journals such as *Advanced Materials, Optical Letters,* and *Applied Optics* but also in biological ones followed, in addition to patents covering production methods of the material, mutant proteins allowing the usage of cheaper lasers, and suggested applications of BR films.[443]

Similar to biotech enthusiasts such as Conrad and Ulmer who regarded genetic engineering as a pathway toward evolving technologies, the German researchers argued that this approach "mark[s] the strategy for the construction of artificial molecules."[444] Around 1990, it may have looked for a moment as if the project actually had a chance of reaching the market: Wacker's business report for 1990 envisaged a prototype for the following year, and the company seems to have been offering the films on a small scale.[445]

This was also the moment when the press began reporting enthusiastically on the novel technology from Munich that relied on "millions of years of highly successful R&D by nature." The German weekly newsmagazine *Der Spiegel* saw researchers "get a step closer to their distant aim, the compact biochip-based supercomputer," and in 1993, Hampp and Oesterhelt received a research prize endowed by a tobacco company for what was baptized "bionics on the molecular scale."[446] These articles not only show a second pitch of the biochip fever around 1990, but also illustrate a change of language: As much as the terminology of the Munich project in the company's reports, patents, or the peer-reviewed journal articles differed from that of *Omni* and the MED proceedings in being much more sober and technical, as soon as the work was taken up by the press, the bold and speculative language from ME and nanotechnology surfaced also here.

Visioneering versus upscaling—materializations of molecular devices

In addition to project organization, type of researcher, and language, there was another central difference between projects to use BR among the MED community and in Munich. This relates to how the actors dealt with the problem of the biochip's materiality, or, put simply, how they addressed the issue of getting the stuff that their chips should be made of. This issue is illuminative with regard to how the more mainstream academic actors from Munich perceived their visionary colleagues from ME, most of them in the US.

Wacker's task in the collaborative Munich project was to develop production methods for BR. This was mundane, but not trivial: Even if research since the 1970s had produced a host of protocols for growing the microbes and preparing the substance, all of these remained on the scale of a research laboratory, and the amounts obtained were negligible if compared to biotechnological production. Moreover, it was not possible to simply make vessels and volumes bigger—fermentation culture was a highly scale-sensitive process (as microbiologists had known since the debut of penicillin production during World War II), and batch preparation (one vessel after another) of a membrane protein could not easily be turned into an economically much more profitable continuous (flow-through) process.[447]

Wacker's patent from 1989 covered a method to produce milligram amounts of the substance. By contrast, publications from the ME field such as Felix Hong's papers proposing the protein as a prototype molecular computing element relied on micrograms only—1,000-fold less. Matters of scale are, of course, relative, and what may have appeared as substance available in boundless quantities to biochemists or biophysicists (a few micrograms were indeed sufficient for sophisticated experiments such as gel electrophoreses, spectroscopy, or membrane reconstitution) was next to nothing when compared to the needs of industrial production. More interestingly, the fact that ME protagonists, many of whom were (bio)physicists or computing experts, did not touch on the issue of production in their publications, which was central to the project involving the chemical industry, is indicative of what these different actors perceived as crucial for a realization of biochips, and possibly of novel technologies in general: Whereas for the ME crowd, the design of a device on paper or a proof-of-principle set-up on a lab scale was central to demonstrate its possibility, a feasible and economic way of producing the substance was at least as important in the Munich project. Put differently, ME focused on biotech as novel, lifelike technology, whereas the mundane but sophisticated aspect of it as a way of producing substances from organisms was not on their map. By contrast, the significance attributed to fermentation culture by Wacker ties in nicely to the pivotal importance of upscaling crystalline silicon production in the postwar development of the American semiconductor industry.[448]

Differing expectations, perceptions, and styles of communication between Hampp, Oesterhelt, and many others based in conventional biochemistry or biophysics in the UK, the US, and the USSR versus novel tech enthusiasts who adopted the formers' object of study prevailed throughout the 1980s. Hampp conceded that the decade was indeed the age of a gold rush (*Goldsucherzeiten*) for ME and early nanotechnology, with everybody going for the big nugget and only time and experience revealing the difficulties and the complexity of turning molecules into working technologies. Yet, at the same time, he was very critical of his ME counterparts, characterizing, e.g., the US physicist Robert Birge (an MED regular publishing on biological computing elements from the 1980s to the mid-1990s) as a "good salesman." [449] Doubting the scientific soundness of Birge's proposals, he was said to have gnashed his teeth when confronted with ideas such as a BR-based 3D-RAM memory. Although these two individuals shared the goal of using this material to build a biochip, for Hampp (and for Oesterhelt) ME had indeed a "louche reputation."[450] Hampp regarded articles on technological usages of the switch protein in popularizing or journalistic outlets such as *Scientific American* (one could add *American Scientist* or *Computer*) as dubious. His main point of critique, notably, was not that these journals were not peer-reviewed, but that they did not contain proper "Materials and Methods" sections, wherein the ways to produce the substance, the amounts obtained, or the technical details of what had actually been built were specified. He perceived it as an achievement that a *Biophysical Journal* paper he coauthored on the use of BR-films as "holographic media" included such data for the first time—to him, this was the first "hard fact" publication on the matter. Indeed, this paper contained data on materials and methods, if also not too many (they may have been part of its commercial side). Interestingly, however, this paper drew in a certain way on ME's suggestive terminology and publications, quoting Birge's, Hong's, and others' suggestions to use the protein as a "molecular switch" in "'biochips.'"[451]

Let it be said, however, that similar opposing views on what were the important steps toward molecular technology, as well as what was a legitimately scientific way of communicating research to peers or the outside, had surfaced already at the first MED meeting in the US, when biochip enthusiast Kevin Ulmer was confronted with what other peers considered simple but crucial counterarguments (the biochip simply rotting away) or unresolved principal matters in the way of molecular technology (this pertained especially to Drexler's molecular assemblers).[452]

As research on BR-based technologies was also carried out in the Soviet Union at the time, debates about the miraculous potentials of this active substance escalated into stories with a Cold War flavor. Under the headline "Vision chemical is found to absorb radar," the venerable *New York Times* reported, for example, in 1987 that Birge's research could "foil Moscow's latest electronic defenses" within three years of an intensive program creating "a new paint, based on chemicals similar to those in the eye, that would render an aircraft, missile, ship or tank virtually invisible to the most advanced radar system."[453] Other bizarre war and spy-tech stories involved pilot's goggles changing color upon illumination, thereby protecting pilots against radiation flashes, or even a BR-based stealth cover for secret agents.

Birge and the Soviet biophysicist Nikolai N. Vsevolodov (then based at an USSR Academy of Science Institute for Molecular Biological Research in Pushchino near Moscow) seem to have been engaged in a reciprocal escalation of what would become possible to achieve with biomolecular technologies. This spiral continued even after the end of the Cold War, when Birge wrote in the *Scientific American* of 1995 that Yuri A. Ovchinnikov, the academician and political functionary who formerly directed the Soviet rhodopsin project, convinced the military that by "exploring bioelectronics, Soviet science could leapfrog the West in computer technology. Many aspects of this ambitious project are still considered military secrets and may never be revealed. [. . .] The details of their most impressive accomplishment, a processor for military radar, remain obscure."[454] Three years later Vsevolodov, who had moved to the US in the 1990s, quoted pieces of Birge's report about what must have been close to him or his own work in a scientific monograph. Thus, he underlined the significance and the hidden potentials of these technologies, without, however, giving much more information on them.[455]

These stories of biomolecular technology's drastically exaggerated potentials seem to have become "folk" among the research community, known by many, and immediately ridiculed or deconstructed by basic scientific considerations in conversation with the author. Some interpreted the feverish escalation of technological ideas and plans as being grounded in cultural differences between US and European science, that is, as deliberate American pitches to receive attention and thereby funding, which probably is not the full story, however, as Soviet sources indicate similar projects and rhetoric. Hampp remembered a Soviet colleague to have been baffled about the milligram quantities of BR he saw when visiting the Munich lab. That said, he spoke quite favorably about the scientific abilities of his Soviet colleagues and the "visionary" aspects of their work.[456]

As unrealistic as these stories (similar to the bio-technological musings of Hameroff or Margulis) may sound in retrospect, they may be best understood as medial resonances of the promise and fascination of novel molecular technologies in the 1980s. Then, a technology that took its inspiration from life promised to be not only smaller, but smarter—one step ahead of others and a panacea to all kinds of economic, social, or even political problems. Generally, this motive has not only been characteristic to debates about nanotechnology ever since, but it also mirrors the "enduring dream" of biotechnologies, promising to improve technology and society by the potentials of life, a leitmotif of this field that has surfaced repeatedly throughout the twentieth century. More recent instantiations of this dream may sound more familiar—think of bionanotechnologies or current media coverage of optogenetics as described at the beginning of this chapter.[457]

This section has also revealed, however, that even if actors in different countries and contexts shared a comparable goal or even vision of using biomaterials for computing, and even if they have been quoting each other's work and texts, their assessments of how to achieve their goals and how to communicate them were not identical. The most important difference could be pinpointed to visioneering versus upscaling—i.e., taking devices sketched on paper as a feasibility proof versus establishing an economically viable technological infrastructure for their realization. And yet, there is a paradoxical twist to this latter approach, which may appear more sound or rational on the first glance: The upscaling of BR production in Munich proceeded further even after the termination of Wacker's project in the early 1990s, with a start-up company by the name of "Munich Innovative Biomaterials," in which both Hampp and Oesterhelt were involved, producing the active matter on the kilogram scale—a trillionfold increase when compared to what researchers had held in their test tubes in the early 1980s. And yet, the smart substance, containing now genetically engineered "molecular switches," is still waiting to be used.[458]

Conclusion I: Assemblers, Cartesian molecular machines, and active matter

The question of why different camps interested in biological technology have zoomed in on the same material in order to make it real clearly has to do with research practice: BR became attractive for the ME community in search of a substrate for their ideas as well as for the Munich project looking for a doable optical biochip for similar reasons that had attracted membranologists in the 1970s—the protein was available as a stable, tractable, and abundant substance.[459]

Yet, this choice may also tell us something about the specific machine concept endorsed by many of those attempting to engineer proteins as molecular "pumps," "switches," or other machinery. Based on what has been stated in chapters 2 and 3, the machines that biochemists and biophysicists from the rhodopsin camp were tackling are best described as Cartesian mechanical devices, i.e., proteins as composed of at least in principle clearly delineable moving parts (hinges, gates, etc.) that accomplished function through rearrangements. Related Cartesian machine concepts were and still are used in research on other models of protein machinery, such as the molecular "rotor" of the ATP-synthase, the myosin protein catalyzing muscle fiber motion through a sliding action, or the ion pumps of the nerve membrane.[460]

A machine concept centering on movement and mechanical causation contrasts to Eric Drexler's assemblers: Remember that in his vision, the existing protein machinery of cells would be able to guide the creation of ever smaller and more refined machinery, eventually from other materials. In other words, assemblers—with the term, Drexler played on an analogy to the computing process of translating a program into machine language—would be machines producing other machines. However, an ability of machines to reproduce (unrealized for all extant devices, apart from software) has been a main ontological point of distinction from organisms since the days of Kant. Reproduction and self-assembly have remained a matter of controversy about the possibility of Drexler-type molecular machines, with many renowned scientists arguing against it.[461] Biochemical and biophysical studies analyzed in this book, however, did not attempt to find ways to construct molecular machines de novo, or to use them as molecular tools to build even smaller or more sophisticated machinery. They had cells synthesize their pumps, channels, or motors, and since they did not primarily attempt to pass beyond the level of miniaturization provided by life, or to conceive of generative machinery, they merely conceived of proteins as Cartesian molecular-mechanical technologies that could be arranged and used as they existed. A review of Felix Hong's 1989 book on biocomputers and biosensors summarizes this distinction between assemblers versus existing molecular-mechanical model systems concisely. Juxtaposing ordered macromolecular arrangements such as Kuhn's LB-films (or the purple membrane) with "a self-organizing molecular machine capable of its self-replication," the reviewing Japanese researcher assumed it would "soon become evident that such an enterprise would be neither easy nor feasible in the foreseeable future."[462]

Distinguishing Cartesian machines, e.g. ordered arrangements of macromolecules that many protein researchers endorsed, from Drexler's generative machines is important: It was this first concept of machinery that rose to prominence in the life sciences after 1990, as shown in the Introduction, not one inspired by assemblers, and it is this concept that abounds in much of today's discourse on protein machines in pharmacology, bionanotechnology, etc. These machines can be decomposed analytically into functional parts, or adapted by genetic engineering, but they remain ultimately designed by natural selection's tinkering, which is a central point of distinction from man-made nanotechnology. However, my supposition that this Cartesian concept of machine is widespread and had a significant impact on research does not imply that it is free of contradictions from a philosophical point of view, which, however may not have mattered so much for the actors. Shortcomings of this concept at the molecular level, and reevaluations in light of current research, will be discussed in the Conclusion.

Against the background of the sweeping organicist discourse of life as an inspiration for alternative technologies in the 1980s, there is some irony to the fact that, of all things biological, a protein that had become a model system for a molecular-mechanical approach to biology, served as a material nucleus for lifelike devices. BR was a special case as it was less a typically delicate biological material than a relatively stable chemical substance that could, e.g., be taken out of its wet cellular milieu and inserted into a synthetic film without disintegrating. The plug-and-play approach offered by this molecule paradoxically turned it into a focal point for those interested in life's complexity or the approximate nature of cognition. One may understand this as simply a case of false-labeling on the part of the actors playing the "bio" card at a time when this was favorable (whereas what they were actually doing was much more physics or chemistry), or else as a methodological consequence of attempts to tackle life by the methods of the late twentieth-century molecular sciences.

Yet, I propose a different interpretation: As described in the Introduction, since the 1990s, many more proteins, from those of our own bodies to those of plants, animals, and microbes, have become amenable to technologies pioneered with BR and other model proteins, and a related mechanical understanding of their functioning has become predominant. So, I argue that taking the substance BR as an exemplar for what "life" represented and how it was to be explained, reveals a more general trend: Research on simple model proteins such as this one has helped to transform the concept of proteins at the molecular level, thereby changing what "biological" or "lifelike" referred to in these respects. Or, to reemphasize a central message of this book: Life has been made mechanical at the molecular level by zooming in on objects that may have actually been as much chemical as biological.[463] To overgeneralize quite a bit: Life isn't what it used to be because of research on objects such as this.

Wherever future conjunctures between technology and life, e.g., in optogenetics, may lead, proteins, their structure, their dynamics, and possibly also their self-assembly will presumably play an important role. To current researchers, and here lie important points of distinction to prior concepts of matter, these complex molecules shaped by evolution represent an active type of matter. Such "smart materials," as Evelyn Fox Keller has put it, are endowed to perform specific physico-chemical processes through their differentiated structure, thereby standing in stark contrast to a conception of matter as homogeneous and passive (see Conclusion).[464] Taken out of the organism and reintegrated into it by means of recombinant DNA and cell technologies, these active materials have the potential to reshape life, technologies, and the relation between them, such as when a beam of light flips a switch in an animal's brain. Thus, proteins as active matter can make us aware that, in unexpected ways, some machines have become more lively, or biological, than we may have thought, and that in turn, life has become more chemical than we have noticed.

Conclusion II: After the fever pitch—a more inclusive history of biotechnology

It is obviously much more difficult to trace the ends of abandoned research projects than those of successful, or at least consequential, ones. Norbert Hampp framed the end point of the Munich biochip project with the disappointed words "here's a great solution, if someone only had a problem for it."[465] Therewith, he was referring to the fact that the coherent laser set-up needed for the image correlator made the device very sophisticated, expensive, and, worse, immobile. In conjunction with changes of personnel and the R&D strategy at Wacker Chemie in the early 1990s, as well as a termination of the grant from the Federal biotech program, the award-winning biochip project was abandoned. This happened at a time when the general problems and the outlook of microelectronics changed: The problem of high-density optical data storage was successfully accomplished using synthetic organic dyes in compact discs, and another impending doom to miniaturization of microelectronics, as prophesied by Moore's law for three decades, did not come true as novel lithographic technologies overcame the limits of the 1980s.[466]

So what historiographical lessons can be drawn from this abandoned project, and from biocomputing in general? First, this story of attempts to turn life's molecular machinery into technologies has revealed that molecular biotech consisted of more than pharma and biomedicine. That the history of the life sciences has analyzed the 1980s mostly through the looking glass of developments that appear as forerunners of today's towering technologies and insights—such as DNA sequencing and genomics, PCR, computer databases, or the commercially successful biomedical face of biotechnology—may have led to the impression that this decade, a generation past by now, was something like an appendix to the present. By contrast, this chapter inserts the 1980s into the *longue durée* of twentieth-century biotechnology broadly construed, that is, as efforts to make use of life in order to create novel, different, even alternative technologies not only to make new products, but to change the human condition.[467] The idea of "molecular bionics," pursued by or argued for by various actors from the life and the physical sciences in this chapter, from visioneers, more conventional academe, or researchers with a political agenda and countercultural background such as Lynn Margulis, shows that 1980s biotech beyond the stories of start-ups, venture capital, and the entrepreneurial scientist existed and that it was a colorful episode of scientific and technological planning—one among many motivations was "better" because organic, more human, smarter (or for others, greener) technology. For the history of molecular machines, this meant a transformation not only of the goals, but of the style and the environments of research. Metaphorically, one could say that while in the 1970s, molecular machinery was a topic in labs in California (such as Stoeckenius' at UCSF), during the 1980s it became a topic of Californian labs, that is, labs dabbling in development of novel technologies for various motivations (possibly even in the Soviet Union). Such a more inclusive version of bio-tech, hyphenated to mark its distance from the narrower project that we mostly remember today, involved specific venues such as the MED meetings, or different scientific media than analyzed in prior chapters of this book—popularizing or generalizing journals and monographs, or the daily press (presumably similar could be said of nano-technology, as Stuart Hameroff spelled it initially). Thus, the 1980s transported discourse about molecular machinery from specialized scientific journals into the public—or, from the *Journal of Biological Chemistry* into *Omni*. This helped shape and spread a technology-centered, and visionary language that has stuck with the field ever since.

By focusing on a project to actually make "biochips," this chapter has contrasted the visionary aspect of biotech and ME with the mundane world of a state-funded R&D program in West Germany, underlining thereby the many roots of the projects labeled today as bio- and nano-technologies in visionary projecting, industrial materials research, or instrument development.[468] Whereas the Munich project illustrates that ideas similar to those heard at the MED meetings were pursued under very different circumstances, my analysis of it has also highlighted the contrasts between what various actors considered as proper science and a proper language of talking about it as they entered technology-centric research. Materiality and materializations of molecular machinery have formed a central point of distinction between the two camps—whereas MED researchers largely did not bother about the problems of making the stuff their chips should be produced from, this mundane but tricky aspect of biotechnology was pivotal to the Munich project.

Finally, this chapter has exposed how 1980s biotech has introduced a different temporality into research on molecular machines. The expectation that molecules would soon be turned into "ultimate" technologies was uttered by a number of heterogeneous actors as their space of technological experience had been dramatically enlarged after the first rapid successes of recombinant DNA from c. 1975–1980. The perceived tempo of research seemed to accelerate in microelectronics and in biotech, with technological breakthroughs appearing as immediate. The impression of coming revolutionary change contrasts not only to the abandonment of the Munich project, but also to the general outcome of molecular electronics' second wave, with research activity going down after 1990, before intensity and expectations went up again with the nanotechnology wave after 2000.[469]

As we know, neither the biochip nor Drexler's assemblers have materialized as technologies so far, and the range of working nanotechnologies seems rather limited. Yet, I suggest it would be too easy and rash to judge efforts of turning molecules into technologies in hindsight as unjustified speculation or as a failure. A more nuanced historical assessment of the 1980s must investigate the distinctions between our actors' horizon of expectation and the present one in more detail: This chapter's actors had relatively little historical experience of recombinant DNA's rapid development, with the approach acting in many cases more like a ferment to raise expectations than actually delivering. This, in combination with a far greater belief in technological (as well as, for some, social and political) change, may explain why established researchers, including not only Britton Chance, Hans Kuhn, or Dieter Oesterhelt, but also Lynn Margulis and many others, believed in the potentials of biocomputing. Our experience of this field as readers of the early twenty-first century, by contrast, is informed by developments over a much longer period, which are characterized not by breakthroughs, but by piecemeal change over decades, a normalization of biotechnology's pace, its integration into existing

pharmaceutical or chemical industries, as well as a number of disappoint-
ments (such as the outcome of the Human Genome Project around 2000,
which did not lead to an immediate revolution in genetic medicine, but
rather complexified matters).[470] Neither have the utopian narratives of
molecular technologies come true, nor have their dystopian counterparts
such as the "grey goo," that is, the scenario of self-replicating molecular
machinery wreaking havoc to the planet.[471] Arguing with Reinhart Koselleck, one may say that this richer space of experience has had an effect
on the early twenty-first-century horizon of expectation—with it becoming, according to him, more circumspect, but also more open.[472]

These differing experiences and outlooks distinguishing present mental maps from those of this chapter's actors may also explain why the
mix of visioneering and mainstream science and technology appears as
such a puzzling, inextricably colorful blend. This probably looked less
astonishing in the gold rush era of molecular technology. In that sense,
the story of biocomputing, as contemporary as it may sound with current expectations about optogenetics resounding in our ears, illustrates
an age of science that is long gone. The seeming paradox of why hopes
for molecular devices could serve as a common vision to a very diverse
range of actors, and blend in an estranging way what neatly fell into the
separate boxes of fact and fiction later on, should be read as a *memento*
for the fact that the time of biotech's hot pursuit almost forty years ago
belongs deeper into the past than we may be accustomed to think. If
the story of biocomputing, awkward as it appears today, helps to produce an alienating effect in perceiving and understanding the 1980s, that
marks their distance to the present as a foreign country, that is already
one good reason to tell it. And if biocomputing casts a new light from
history on current fever pitches such as in optogenetics, where certain
tropes of revolutionary change in the relationship between organisms and
technology have reappeared in a different guise, that may be another one.

Conclusion

Toward the end of his 1970 *Logic of Life*, François Jacob uttered that in molecular biological laboratories of his day, "life" was not questioned any longer. Published only four years after Michel Foucault had famously historicized the concepts of life and biology as bound to a modern "order of things," Jacob's conjecture was motivated by postwar molecular genetics, understood not only as an inquiry of organisms or heredity, but more generally as one of code and information transfer under the umbrella of cybernetics. However, the last chapter of Jacob's book, a big picture of organizational steps in "living systems" from atomic structure and molecular self-organization to culture and societies (NB, published at the time when this book's narrative set in), hints at another answer to what the life sciences may explain at their lowest level in lieu of life, namely, the intrinsic properties of matter that lead to the formation of complex structures. Modern biology, thus Jacob in another programmatic statement at the outset of his book, explains the organism from the structure of the molecules constituting it, and in this sense it corresponds to a new era of mechanism.[473]

So, if recent biological research does not question life any longer, in studies of membranes and molecular machinery from microbes to human beings, of molecules that rotate, twist and bend, dock, push, and pull, with pieces and parts binding, fusing or falling apart, this research questions "matter." Specific forms of matter are acted upon if our contemporaries ingest a proton pump inhibitor pill to alleviate the symptoms of heartburn by blocking a pump in their gastric mucosa, or if optogenetics researchers shine a light on a neuronal membrane to open the ion channels therein. In this section, I will analyze and contextualize the turn to matter as the lowest level of analysis of life, or more specifically speaking the composition, structure, and activity of specific forms of matter allowing the contemporary sciences to explain biological phenomena in the form of molecular mechanisms. The resulting molecular-mechanical vision of life, as conveyed by the plethora of books, journal articles, or advertisements discussed above, thus represents much more than a strategy or simplification to communicate scientific knowledge. Rather, it should be considered as a full-blown research program at the cross-roads of chemistry, physics, and the biological sciences that has transformed the materiality of life in the past four decades.

I will first return to the central epistemological issue raised by this book, namely what a material-centered, historical approach such as this one can add to the so far mostly philosophical debate about molecular mechanisms, and to what extent it allows a qualified answer to the vexing question of what molecular machines have represented to scientists, as concepts with explanatory value and a material reference—should we indeed take them for real? The two following sections of this Conclusion are historiographical: I will zoom in on the development of the twentieth-century life sciences at large, asking to what extent the present story allows us to expand and reposition the received picture of life's molecularization, especially with respect to its relationship to practices and concepts from chemistry. Finally, I will raise some issues pertaining to this book's geography and actors. The question of what type of scientist we have encountered in this book may inspire further inquiries into recent science.

Matter, activity, and mechanisms at the interstice of the chemical and the life sciences

The red thread structuring this book's narrative from the 1960s to around 1990 has been formed by researchers' observations and interactions with active matter, and its subsequent materializations as molecular machinery. This led from the encounter of a curious membrane preparation that changed its color to the isolation of a molecular pump, the dynamics of which caught the attention of biophysicists interested in molecular mechanisms, as much as of synthetic biologists avant la lettre establishing a "plug-and-play" mode of experimenting with life's material inventory. Finally, the potential and promise of this active matter for molecular devices entranced bio- and nanotech advocates in the 1980s: Here was a substance that could be obtained as bulk matter in the thick, composed of molecules that promised to perform certain "jobs" such as pumping or switching with high precision—thus, a well-suited and tractable materialization of molecular devices.

As noted in chapter 2, the German concept *Stoff* seems particularly apt to grasp the macroscopic, perceivable aspect of materials that has been important for this book's materiality-centered historiography. *Stoff*, translated at a certain loss into "material substance," refers not only to chemical, but also to textile matter, that is, to produced, tangibly structured, and usable things that are not found in clearly discernible objects or units. Sulfur or its composites, for example, are *Stoffe* (plural) endowed with certain properties, such as their composition, molecular structure, or reactivity.[474] The direct observations and manipulations documented in laboratory notebooks, images, and technical parts of publications (such as "Materials and Methods" sections) that have informed this book's narrative bring into view the often quotidian aspects of laboratory work with Stoff. In other words, the material and mundane side of research illustrate how far the specific (re)activities of matter, or its agency in a sense compatible with modern chemistry, have influenced the development of inquiries. In this story, for example, material encounters have been influential at important turning points of the researchers' intentions (such as from the study of membranes to that of a specific molecule, or toward the novel method of crystallographic EM; see chapter 2). In turn, such encounters have engendered novel materializations of molecular machinery, with the latter becoming more articulate as a preparation or as molecules of known composition and structure.

Particularly in my reconstruction of 1970s membrane research from notebooks, aspects of materiality have frequently been documented not solely through data or traces recorded by instruments, but also through simple tinkering with Stoff. Often enough, materials existed as perceptible matter in bulk, displaying visible effects, such as changing color, precipitating, or even disintegrating in certain chemical routines. That is, my analysis at the level of preparatory work and routines has revealed the relevance of a tangible dimension of material substances that may be unexpected for the late twentieth-century life and chemical sciences, which are, at least since Bruno Latour and Steve Woolgar's *Laboratory Life* (1979), better known for sophisticated instruments (such as spectrometers or X-ray machines) analyzing traces of materials, and producing inscriptions (graphs, photographical patterns, etc.).[475] By contrast, many material interactions described here are more reminiscent of earlier times in chemistry's history, when materials displayed their properties and effects in the most obtrusive ways—as dirt, smell, or even through explosions—with materials being not elusive and hard-won as proper scientific objects, but sometimes as evident as a slap in the face.[476] This untimely resemblance between materiality in the molecular life sciences and a chemical laboratory from the early twentieth century or before should not be considered as historical revisionism—not least, modern physical instruments and their data were equally important in this story. However, it may serve as a reminder of the role of chemical thinking and working in the life sciences generally: On the level of tinkering with material, perceptible qualities, the tangible aspect of matter (its molar aspect as opposed to its molecular microstructure), still play a significant role. To take a "feeling for Stoff" (paraphrasing Evelyn Fox Keller) on the part of the scientists into account, material-centered histories have to look beyond the technological feats that appear prominently in papers and recollections of researchers, and focus instead on sometimes tedious preparatory work, as documented in protocols or notebooks. Analyzing the development of this mundane dimension of work with materials over time could also reveal different temporalities of research practice—with the "old" simple tinkering experiment and routine work not disappearing, but persisting alongside new high-tech research. Following Steven Shapin's argument about the invisibility of preparatory and manual labor, one may assume that such material aspects of research were relegated to technicians or PhD students, and that they may be documented predominantly in unpublished records.[477] Still, even in the age of X-ray or sequencing machines, somebody must find his or her way through a maze of liquids, greases, and precipitates, somebody has to have the skills to interact with materials directly, and observe properties and effects with their eyes and noses. In this story, and probably other cases could be found, it was in the course of such labors that a Stoff with very specific properties and activities was encountered, forming the nucleus of a new research project. Consequently, the chic and slick molecular objects depicted on today's journal covers, in advertisements, or in books such as Goodsell's exhibit ineffaceable continuities with the realm of direct perception and manual action: Life as matter unmade and remade by the human hand and mind.

The relevance of materializations for the unfolding of this story, and arguably that of molecular machinery at large, begs for a positioning of active biological matter broadly construed within the history of the twentieth-century life sciences. First, the assignment of biological functions to the properties of specific biochemical substances is not at all new—rather, this development is neatly tied into the history of biochemistry: The isolation of enzymes in the early twentieth century, for example, which were able to perform feats such as the conversion of sugar into alcohol in a test tube, allowed early biochemists to map the biological process of fermentation on the specific reactivity of a material substance. With the help of instruments such as the ultracentrifuge and the electron microscope, viruses materialized as chemical objects since the 1930s, thereby turning into models to illustrate specific properties of life such as its self-assembly and reproduction.[478]

Yet, with molecular composition, structure, and reactivities of proteins spelled out as molecular mechanisms since the 1970s, this process of materialization, and possibly of a "chemicalization," of biological processes gained new momentum: Encompassing the molecular machinery (pumps, channels, motors, etc.) described in this book as well as other proteins catalyzing, e.g., motions of the muscle fibers, or perceiving and self-assembling molecules, the concept of specifically structured and reactive material substances has formed an umbrella to explain living processes within a general physico-chemical theory of matter. As Evelyn Fox Keller and Bernadette Bensaude-Vincent have noted, such active materials, biological or not, have challenged the often gendered passivity inherent in many earlier concepts of matter, especially mechanistic ones since Descartes, and they have brought to the fore the relevance of chemical thinking and working. Chemistry has provided concepts to accommodate the specificity and the activity of material processes with a naturalistic understanding of life. Moreover, and here we reencounter Jacob's conjecture quoted at the beginning, the increasing relevance of examples such as the self-assembly of virus particles or the dynamics of protein machinery within the recent life sciences has the potential to yet again blur the boundary between living and non-enlivened, which seemed relatively clear-cut for much of the twentieth century.[479]

Put briefly: The modern chemical concept of matter, encompassing its potential for self-organization or specific reactions, which are inherent in today's molecular-mechanical perspective on life, mark a central difference to the early modern mechanist conception of living processes. Moreover, what chemistry tells us about reactions also helps to disentangle active matter from any sort of "molecular vitalism" or even hylozoism, that is, of specific nonphysical forces animating matter, which have informed discourse on the subject for long time, and which seem to be lurking at the back door of contemporary debates. As strange and entrancing as the phenomena of self-organizing, reacting, or "excitable" matter may appear to outside observers and researchers alike, these should not be taken as indicative of biomolecules behaving in principle differently than their inorganic counterparts, let alone of them alone being "alive." Active matter is firmly grounded within a physico-chemical framework of explanation describing such phenomena by reaction kinetics, i.e., as enthalpic and entropic processes following the laws of thermodynamics.[480]

On the basis of the present story, I argue that zooming in on active matter, as a recent umbrella concept with a far-reaching genealogy in biochemistry and biophysics, reveals a transformation of the ontology of organisms, which are now conceived of as composed by specifically structured and reacting molecular components. In turn, focusing on biological cases of active matter also illustrates the concurrent change of the concept of a chemical substance. Genetically engineered proteins as dynamic molecules that can be shuffled between organisms, that are used as commodities (enzymes in food, pharmaceuticals, or washing powder; chip technologies; bionanotechnologies), that can be applied to other organisms in cancer therapies (e.g., antibodies) or even integrated into living tissues in optogenetics and implantation technologies, form a group of complex substances that was, if not unknown, then only of tangential importance to research and industry prior to the 1980s, or even the more recent past. Such complex macromolecular substances, which biotech or chemical companies were keen to add to their portfolios, often display different properties from the conventional (bio)organic substances with which chemists have been acquainted, say alcohols, fats, or carbohydrates: These novel substances frequently exist in small amounts only, as they are difficult to isolate or synthesize (e.g., by biotechnological routines), they are often delicate with respect to their conditions of existence (biological matter, for example, needs to be kept sterile, cooled, etc.), and they display properties and activities sensitive to their milieu, such as photoreactions or electrical effects, changes of transparency, contraction and expansion, or filtering.[481]

It is the increasing availability, study, and relevance of this type of active, complex matter in science, medicine, and technology that has formed the basis as well as the infrastructure for the rise of the molecular-mechanical vision in recent decades. That is, not only were biological functions pinpointed to proteins as chemical substances to an ever increasing degree, but these substances were increasingly made and unmade using the methods of organic chemistry and recombinant DNA, analyzed and modeled in plasticine and wood, on paper, or on screens as structured entities reacting and moving in space. They have populated a novel cosmos at the interstice of life and matter that is with us today far beyond research—in medical practice, advertisements, and consumer culture. To put it differently: Bodies, tissues, and cells of our present appear so self-evidently as filled with a cosmos of active matter, of materialized molecular machinery, which even scholars of science such as Evelyn Fox Keller take as the natural way of how things are, because "life" now has a very different materiality than it had forty years ago.[482] Did these substances not exist prior to c. 1970? Some of them certainly not, such as recombinant proteins, others possibly, albeit in a sense that would need to be defined more clearly in philosophical terms, as they were not in the hands of scientists as isolated material substances, as Stoff on hands, nor were many of them named, analyzed, or represented as discrete objects. That is, if they were part of the laboratory ontology, then in a different way.

From the perspective I am suggesting here, the molecular life sciences appear to be not only an "engine of discovery," as Lindley Darden and Carl Craver have put it in their recent book, and as the philosophical debate about mechanisms seems to assume widely.[483] Rather, in the view I propose as a necessary historical complement that takes into view practice and materiality of research, these sciences have to be considered on equal terms as a human activity that tackles organisms and their material components, taking them to pieces and remaking them, and thereby shifting the identities of both the living and the chemical realms. This book and its case story should be considered only a starting point for further historical as well as philosophical studies on recent changes of life's materiality.

From what has been discussed above, it appears clear that the age-old problem of the relationship between machines, mechanisms, and organisms must have found a new guise in the age of active matter and molecular machinery, and it may have become a different problem altogether. On the one hand, the problem as it can be gleaned from this story has shifted from that whether organisms are, or can be represented as, machine-like entities (dealt with by Descartes, Kant, or Canguilhem) to that if their molecular components represented some sort of machinery.[484] Arguably, the materializations of this machinery, their conceptualization as active matter in terms of chemistry, and the ability to act on them, e.g., by pharmaceuticals—if you can block them, they must be real—mark important points of distinction to earlier theories or models of organisms as mechanical devices.

However, there seems to be a common blind spot to machine analogies to organisms as well as to molecules. As discussed in the Introduction, a central argument to the work of Canguilhem, which can be traced back to Kant's characterization of organisms in the *Critique of Judgment*, states that mechanisms may tell us something about the working of a machine, but nothing on how to build it, and that therefore mechanistic explanations fail when it comes to organismic reproduction. This restricted purview of mechanistic explanations maps on the machine concepts largely operative in the discourse studied in this book: The specific concept of molecular machinery advocated by biochemists and biophysicists and by a number of actors in the biocomputing field was that of Cartesian mechanical devices, with function being accomplished by defined motions in space (albeit based on a different conception of specific, reactive matter). By contrast, Eric Drexler's "molecular assemblers," which most advocates of protein molecular machinery analyzed here were critical of or remained mute about, were to self-reproduce, thereby becoming truly lifelike machines.

In other words, the possibility of molecular machines' reproduction, a pivotal philosophical point of distinction between organisms and machines, seems not to have been central for many biochemists and biophysicists working on molecular machinery in the 1970s and 1980s, and possibly also later on (see chapter 4). From an entirely historicist point of view, one has to concede that the neglect of this philosophical problem has not hampered the rise of the molecular-mechanical vision of life or the productivity of the machine analogy in science so far. From a philosophical point of view, however, one may suspect that this discourse has not solved the riddle of these machines' design and reproduction—the machinery discussed here has been produced by cells, and thereby organism-like entities, with the "designer" being evolution. This leads to the critical question if the cell would not have to represent more than a mere "collection of protein machines" (to turn around the programmatic statement by Bruce Alberts in 1998). In parallel to Canguilhem's central argument about the epistemic primacy of the organism and the derivative status of machines as built by living beings according to their functions, this would imply that the cell should be considered as prior and primary for the design of molecular-mechanical devices, and not the other way around.[485]

Yet, to me there appear to be alternatives to either such an organicist position, which comes with lots of philosophical baggage, or an entirely historicist position on the matter. To understand what was conceivable for the scientists in the 1970s and 1980s regarding molecular reproduction, it would be necessary to take into account contemporary discourse about self-assembly and self-organization, which can only be hinted at here, as this took place in a different arena. Notably the theories and models of physicists such as Manfred Eigen's autocatalytic "hypercycle," Ilya Prigogine and Isabelle Stenger's books on structure formation and temporality, or scenarios for a molecular origin of life (the RNA world)—largely desiderata of historical scholarship—may have provided a possible implicit background to conceive of evolutionary explanations for how molecular machinery could come into existence—in the history of the earth as well as in test tubes. Donna Haraway, for one, found the distinction between organisms and late twentieth-century machinery "leaky". [486] May this not reflect what François Jacob had stated a decade previously—that "life" as we knew it was not the question any longer?

Molecular machinery in past, present, and beyond

It is time to face a crucial question that may have been on the mind of many readers—after all that has been said, historically and epistemologically, to what extent can we say that proteins actually *are*, or have become, molecular machines since the 1970s? Before answering this question, let me briefly summarize the main premises of this assumption: Proteins are modeled as objects with a clearly defined outline or shape, containing defined elements of substructure, which move in space in order to achieve function by structural rearrangements, i.e., conformational changes, and/or by catalyzing chemical reactions. Function is frequently displayed as a series of cartoon-like snapshots, or nowadays computer animations, of these proteins in different states. The molecular models used in these explanations have resulted to a large degree from X-ray crystallography, though not entirely, whereas the knowledge about protein dynamics is based on spectroscopies; biochemical, physiological, or genetic experimentation; as well as molecular dynamics simulations.

With the story of this book in mind, it should have become more articulate what I meant when paraphrasing Ian Hacking in the Introduction with the line that if molecular machinery can be interfered with, it must be real in some way: This was to say that many actors in the molecular life sciences, and some philosophers or historians, take molecular machinery as more than a tool for pedagogy or a catchy slogan for communication—for them, this concept has found concrete materializations and it has explanatory value. There are many episodes from this book underlining this—from the first materializations of membrane pumps in the 1960s and 1970s, which sparked consequential research, to the channels used by present optogeneticists to manipulate cells and animals, to specific studies taking proteins apart and putting them back together, or inquiring into functional elements explained in mechanical terminology, such as molecular "switches" or "hinges." As a result, concepts such as that of a molecular pump are nowadays used in a self-evident way in the scientific literature, and they have become part of a broader discourse on life, such as illustrated by David Goodsell's book, in advertisements, etc.[487] In other words, molecular machinery as part and parcel of the molecular-mechanical vision of life has received credibility within science and beyond on the basis of transforming life's materiality, by its explanatory value, and finally by its promise to allow manipulations of biological processes.

However, I would like to understand my argument for accepting molecular machinery as part of science's ontology, which resonates with entity realism in the philosophy of science as advocated by Hacking and others, not as a definitive or normative argument for the "reality" of these entities, but as showing how and why scientists have come to regard them as real. To estimate where similar developments of metaphors turning literal and a materialization of entities may lead, think of the cell. As historian of science Andrew Reynolds has argued, this concept, initially used to describe units of bounded space, but now self-evident to biologists, has been based on metaphor since the beginning of its usage in nineteenth-century biology, with shifting references and meanings throughout the development of the life sciences. And who would doubt the existence of cells now?[488]

Time will tell whether future developments in medicine, technology, and science will lead to a similar naturalization of molecular machinery, what revisions the specific machine concept may undergo, what new discursive connections may form or which new materializations may be produced. It is also possible that molecular machinery as outlined here may appear as bound to a state of scientific discourse and practice in the last decades of the twentieth century, and may give way to other concepts.

There are developments in current protein research suggesting that molecular machines and mechanisms as outlined in this book are already being amended and superseded by new insight. First, un-black-boxing of crystallographic practices reveal the limits of the suggestive models of molecules as thing-like objects. Crystals, as used for structure determination of proteins, do not represent the native, more dynamic state of most proteins floating around in a cell or a membrane. Contemporary computer simulations of molecular dynamics, for example, provide insight into the fluctuations and changes of molecular structure over time. What is more, the conventions for graphic modeling of crystallographic data, or how to represent protein substructures, that have been put into place since the 1970s contrast with the individual character of proteins, for example, as some portions of a specific molecule may be more dynamic than others.[489] So, a protein that appears as a rigid structure in the models depicted in this book may be in constant flux in its physiological milieu, i.e., the molecule is always present in a number of interchanging conformational states. Recently, this dynamic perspective on protein function has been supported by spectroscopic evidence, notably from nuclear magnetic resonance (NMR). The advantage of NMR is that proteins are studied in an environment much closer to the physiological one, that is, in watery solution rather than in a crystal, and that the method permits biophysicists to record structural dynamics over time, such as when a substrate of an enzyme is added.[490] NMR models of protein structure have built on X-ray crystallography and at the same time overcome some of its limitations: The basic idea of visualizing protein structure through a spatial arrangement of elements (such as alpha-helices) is similar, as is their graphic display. However, NMR models do not provide a single static image of a molecule, but always a dynamic ensemble of probable states.[491]

The static and the dynamic approach to proteins, represented by crystallography versus spectroscopies, constitute complementary but not necessarily mutually exclusive representations of the molecular world: Whereas some proteins and their functions are still today modeled in a more orthodox mechanical way, with rigid structures and clearly delineated movements, others seem to be more protean—hard-to-grasp, floppy, shifting—obeying the statistical laws of the microworld rather than classical mechanics of macroscopic devices. This latter perspective may not be easily reconcilable with a concept of Cartesian molecular machines.[492] Recent nanotechnologies, setting out to construct molecular devices, have revealed yet other limits of Cartesian models by exploring the differences between mechanics at the macro- and the microscales: Physico-chemical laws play out in other ways for what British chemist Richard L. Jones has called "soft machines," such as when thermal vibration of molecules (Brownian motion) is taken into account, which implies that parts of the machines permanently wobble around. Another feature of the microscale affecting both functioning and assembly of molecular devices is their "stickiness," i.e., the adhesion of molecules to each other by chemical interactions.[493] Exploring such peculiarities of the microscale further, philosopher of science Sacha Loeve has argued that even the problem of a machine's individuation, the relationship of the whole to its parts or that of an object to its environment, calls for different concepts of machines, as well as of objects generally, especially when we think of synthetic nanodevices. These differ from proteins in that they need to be fabricated by humans and not by evolution, and that their usage needs to be controlled, not least for safety.[494]

These and other contemporary debates in the science and technology of macromolecules corroborate the impression of an ongoing development of the machine concept, and possibly a modification or even displacement from the point in the late twentieth century to which I have followed it historically. The outcome of these debates will decide how future historians may conceive of the history of proteins as molecular machines—as a beginning, as one stage in a longer development, or as a dead end.

The bigger picture—membranes and molecular machines in the history of the life and the chemical sciences

My tentative response to François Jacob's conjecture has been that laboratories of membrane and protein research may have questioned matter rather than life in the recent past. Not only have they investigated active, structured matter such as proteins, but they have thereby remade bits and pieces of life. The centrality of Stoff in the life sciences prompts us to question and reevaluate their relationship to chemistry, as the science investigating the properties of matter. Thus, on the basis of this story, we may revisit some long-standing assumptions and narratives of the life science's twentieth-century history, especially with regard to the decades after 1970, that is, after the peak of postwar molecular biology.

It is clear that taking into account the concepts, actors, periodizations, but most of all the topic of the "membrane moment" around 1970 and its aftermath makes the contours of a novel narrative of the life sciences' development discernible: Membrane and protein science as presented here have not set out to provide molecular explanations of organisms' reproduction or heredity (which structured Jacob's account), but of their dynamics, that is, of problems related to energy, metabolism, movement, or cognition. These have been longstanding and intensely investigated topics of the life sciences construed large (i.e., encompassing physiology, biological, organic, or colloidal chemistry), and in fact they have been the subject of a stream of historical investigation before c. 1990, such as in Robert Kohler's work on early twentieth-century biochemistry and the enzyme theory, or Frederic Holmes' studies of chemical investigations into life from Lavoisier and the chemical revolution to Hans Krebs and intermediary metabolism.[495] One reason these large fields of study have faded out of view in recent historiography may be, to simplify quite a bit, that they lacked a Darwin or a Mendel, and they have lacked a Darwin or a Mendel as we know them since they lacked a Watson, a Crick, and a Human Genome Project, an E. O. Wilson or a Richard Dawkins (that said, Peter Mitchell with his strong personality and his philosophical leanings appears as a possible candidate).[496] Or, in Angela Creager's words, physiology or physiological chemistry and what developed out of them in the twentieth century as biochemistry, biophysics, etc., may have not been as visible or obviously politicized domains of the life sciences as genetics.[497]

However, there are reasons intrinsic to science that these approaches to study life are harder to grasp historically: They were much more scattered in between biology, chemistry, and medicine, lacking the conceptual thrust, the disciplinary coherence, and the actor's historiography of twentieth-century genetics.

But does that mean that explaining features of life such as movement, metabolism, or even thought by the properties of organized, active substances is any less relevant for modern societies? Or that this was less of a political project than explaining heredity? A short glimpse into the history of physiology, from inquiries into the "human motor," or along the line of nerve and cognition from Helmholtz to Hodgkin and Huxley's action potential to today's optogenetics shows that this question is rhetorical.[498] Similar could and has been said of theories of metabolism from Justus Liebig's nutritional chemistry in the nineteenth century to chemiosmosis providing a general, physico-chemical explanation of how cells transform food into energy, with ATP as a "currency," to the post-1970s life sciences rendering these processes in terms of molecular machinery that can be moved and blocked. The materialistic worldview that comes with these theories, outspoken or not, but effectively put into practice in medicine, nutrition, etc., has informed and is informed by political convictions of various sorts. This has been shown for Justus Liebig and his adversaries such as Carl Vogt or Jacob Moleschott in nineteenth-century chemistry and physiology, or for debates on the origin of life by what would now be called molecular self-organization through left-leaning biologists of the first half of the century, such as J. B. S. Haldane or Alexander I. Oparin.[499]

If the Rockefeller Foundation's support for a novel molecular biology since the 1920s was linked to the propagation of specific forms of social control, as Lily Kay has argued, why not ask what broader social and political ideas have informed research into the current molecular-mechanical vision of life, as exemplified by membranes and molecular machines, but possibly also many other topics?[500] Clearly, concepts such as that of the "neuromolecular gaze," proposed by Joelle Abi-Rached and Nikolas Rose, or the "somatic self," grasp important aspects of a present constellation that begs for more historical scholarship.[501] For optogenetics, as yet only a research method, but invested with high expectations to cure all kinds of neurological ailments—from "switching off" depression or narcolepsy by modulating the brain's protein machinery to visual prostheses—bodily control is an openly stated aim. Targeting the cell's molecular machinery to achieve control and restore or even optimize certain functions is daily practice around the globe for drug-based therapies of widespread diseases from reflux esophagus to cancer.

In brief, within a bigger historical picture, the molecular-mechanical vision allowing and promising specific interventions into bodily processes, has been growing out of a heterogeneous, but influential stream of research investigating life's material makeup and dynamics at the crossroads of chemistry and physics that existed throughout the twentieth century (as described in chapter 1), and which has continuously been entangled with medical, economic, and sociopolitical ideas of dealing with nutrition, disease, etc. The present study has been primarily occupied by contouring this "molecular biology beyond genetics" as an object of historical study from within science, and it has focused on the problem of materiality. Yet, this book will hopefully serve as a resource and a point of departure to address this topic from other angles in the future. Looking beyond the gene in this way may also help to overcome a paradoxical effect of recent historiography of the life sciences that seems to have reinforced a narrative structured by molecular genetics even as it criticized the latter's epistemological premises, such as when Evelyn Fox Keller has called the last century that of the gene.[502] In turn, this refocusing may help to historicize and contextualize what appears so much as a given in today's research and even the everyday life—that our bodies are composed of molecular machinery.

Which brings me to an epistemological consideration: The case of membranes and molecular machines is also illustrative regarding the question of what became of the reductionist ideal and heritage of molecular biology, i.e., the question as to whether and what biological phenomena can and should be explained in terms of physics and chemistry, and the level of generality that can be achieved, as famously expressed in Jacques Monod's dictum that what was true for *E. coli* was true for the elephant. In recent historiography of the life sciences, one sometimes gets the impression that this ideal and style of explanation was lost in the 1970s, when it became clear that many biological processes were more complex than assumed at the heyday of postwar molecular biology, and that explanations on the level of cells and organisms have had a comeback, particularly in fields such as developmental biology or epigenetics.[503] This is certainly true, but only one side of the coin: The molecular-mechanical vision shows that aspects of molecular biology's heritage, namely a physico-chemical approach and style of explanation for biological phenomena, seems alive and well in other quarters of the life sciences. However, with important differences: proteins have become at least as important as DNA (in fact, they probably have always been), and research today has revealed a great diversity of highly individualized molecular mechanisms, the "Shakespearean" narrative explanations of which contrast to a "Newtonian" biology of general mathematical expressions.[504] That is, whereas the quest for general principles or laws of how organisms work on the molecular level might have been abandoned by many of the actors described in this story (it has reappeared among contemporary systems biology, however), the underlying assumption or idea that life processes can be explained by molecular-mechanical interactions and reactions seems to be widely endorsed (which, of course, does not exclude higher level explanations).[505] However, against the relevance of matter and chemical concepts exposed by this book, would it not be more suitable to call this epistemic framework explaining life instead of reductionism a form of materialism, potentially even a "chemicalism"?

Let us turn from epistemology to historiography. The story of what happened to molecular biology after 1970 has so far mainly been researched along three interconnected lines: First, the turn of molecular biologists to new topics such as higher organisms and thereby development—one may think of François Jacob working on mice, or Sidney Brenner taking up the worm Caenorhabditis at Cambridge's LMB. Second, there was biotech, and molecular biologists becoming involved in biomedicine and entrepreneurship in the wake of recombinant DNA—this is the story of firms such as Genentech, but also that of a reconfiguration of research, to simplify quite a bit, from analyzing to making. The third well-researched aspect is the development of molecular genetics into genomics through sequencing and large-scale collaborative projects such as the Human Genome Project. These developments have been connected to the introduction of computing and databases and the resurgence of a natural history mode of investigation in the molecular life sciences, i.e., of collecting and comparing DNA sequences, protein structure data, etc.[506] The membrane moment around 1970, as analyzed in chapter 1 and 2, which followed on a long stagnation of the field since the interwar period, and contributed to the subsequent unfolding of the molecular-mechanical vision of protein function, adds a novel aspect to this historiographical picture.

The events that made this moment drew on many resources from postwar molecular biology, such as structural biology of DNA and proteins, and later on recombinant DNA as a way to study and to remake proteins, yet these were combined with concepts and approaches from bioenergetics, physiology, biophysics, or organic and colloidal chemistry.[507] As a result, the "molecular life sciences" as they took shape during the 1980s appear as a merger drawing on these different resources, often enough integrating them in projects investigating one specific protein or mechanism (chapter 3). So, this story stands in no contradiction to either the historiography of postwar molecular biology or the post-1970 molecular life sciences, but it complexifies both: Neither has "molecular biology" pre-1970s been as what it largely still appears (i.e., structural biology plus genetics), nor are the molecular life sciences post-1970 only higher organisms, biotech or genomes and computers, to caricature the situation.[508] By focusing on the materiality of research, which includes infrastructures provided by methods and instruments, and thereby conceiving less of a history of research programs than of a genealogy of practices (chapter 3), this book's narrative crisscrosses and undermines the existing historiographical narratives of the molecular life sciences. The ongoing relevance and productivity of "old" methods from organic chemistry, for example, or of the Warburg apparatus from prewar cell physiology in a cutting-edge molecular project of the 1970s, contrasts to narratives centered on innovation and high-tech, such as sequencing machines or computational analyses. This change of perspective serves as an antidote to overfocusing on highly visible centers and actors of research, and it may even be understood as an argument for research policy to take into account the productivity and ongoing importance of hands-on, bench-top research in the light of the "big life sciences," such as the Human Genome Project and its follow-ups.[509] My genealogical account of synthetic biology should be understood in a similar way, revealing that this recent field, in spite of all claims of novelty, has been built on an ongoing, continuous stream of mundane practices to remake not organisms, but first and foremost the Stoff that life is made of, based to no small extent on methods borrowed from colloidal or organic chemistry that were interlaced with recombinant DNA.

Are these examples that shift perspective from large to small, or from innovation to use, sufficient to produce a "shock of the old" in the understanding of recent science, as David Edgerton has demanded for the history of technology? What is certain is that such stories readjust the historical picture of scientific development by bringing into view an infrastructural side of research, which seems to have special connections to chemical practices. Thus, by looking beyond innovation, stories such as the ones told here may also permit historians to question the unfolding of recent science in time, revealing a greater temporal heterogeneity of developments than assumed previously.[510]

Beyond life? Places and scientists after molecular biology

This book has been largely a history of materiality as well as of approaches, instruments, and methods. On the flip side, however, it could also be read as telling us something about the sites of research and the scientific persona in the recent molecular life sciences. To make a full argument of possible developments on these subjects would require other sources and a different historiographical approach. Here, I can only sketch some admittedly speculative thoughts on this topic based on what I have learned from numerous conversations, exchanges of emails and letters, on-site visits in Germany, Britain, and the United States, and not least by reading in between the lines of my sources.

Regarding the geographical distribution of the events covered in this book, one may also understand its narrative as one of a *broadening* of the molecular life sciences. The handful of hotspots at which molecular biology unfolded (in the received sense as molecular genetics plus structural biology) represented the political situation after 1945: In addition to the American sites, such as Boston and Cold Spring Harbor on the East or California on the West Coast, the European centers were located in Paris, Cambridge, and London. Starting in the electron microscopy department of George Palade at Rockefeller University in the 1960s, and bringing in Cambridge's LMB, and to a lesser degree also Maurice Wilkin's biophysics department at King's College, London, this story clearly grew from these places, but it also reveals a more complex geography from the beginning, by highlighting the relevance of the biochemistry of intermediary metabolism, enzymology, and cell physiology (Feodor Lynen's institute at Munich and Warburg's at Berlin), spectroscopy (Stoeckenius' link to photosynthesis through Melvin Calvin at Berkeley, Benno Hess at Dortmund, or Britton Chance at Philadelphia), and bioenergetics (Efraim Racker at Cornell).

Arguably, the most visible change of geographical distribution in this story, however, was the appearance (or return) of other countries on the global map of the molecular life sciences. As Ute Deichmann has argued,

West Germany had largely missed out on the rise of molecular biology after World War II, due to forced emigration, loss of resources, and self-isolation after the war. The new research campus of the MPG, opened in 1973 at Martinsried south of Munich, could be understood as symbolic for a gradual reappearance of this part of the country on the international scene: It brought together a number of still influential important individuals from biochemistry with a controversial past, such as Adolf Butenandt, head of the MPG in the 1960s, with figures that had risen in the postwar period, such as structural biologists and virus researchers, but most importantly for this story Feodor Lynen, one of the few protagonists of dynamic, intermediary metabolism biochemistry left in the country. Lynen became the first postwar full professor of biochemistry in the country as well as a winner of a biochemical Nobel prize.[511] The opening of Martinsried has to be considered in the context of other novel institutions in West Germany, such as the Institute of Genetics at University of Cologne in 1962, which had brought DNA and virus research into the country under the aegis of Max Delbrück, the MPG's Institute of Molecular Genetics opened in 1964 in West Berlin (in immediate vicinity to Warburg's institute), or the decision by John Kendrew to build a transnational center for the molecular life sciences, the European Laboratory of Molecular Biology (EMBL), at Heidelberg.[512] The campus built on the fields south of Munich, also an example for a reorganization and a suburbanization of science, exemplified a new interdisciplinary and integrative approach of the molecular life sciences in the structure of its buildings, composed of a set of concrete blocks loosely gathering around a center, and made to house varying numbers of research groups in flexible proximity. Indeed, Martinsried became a hub for various things molecular biological in the coming decades—from structural biology to molecular genetics or recombinant DNA.

At large, one may tentatively describe the development seen in this book as a dispersion of molecular biology's core (qua molecular genetics and structural biology, as a method and a style of explanation) on the geographical, the institutional, and the epistemic level, such as in its interlacings with organic or biochemistry. And maybe it is this dispersion, which makes the heterogeneous influences of this field from before 1970 as well as before 1945 visible and more interesting historiographically. Consequently, one may ask whether similar developments to the one sketched for West Germany could be found in other parts of the world— the other example that could be touched upon only briefly here was the reinvigoration of the molecular life sciences (including molecular genetics) in the Soviet Union after 1970. These developments obviously took place under very different political circumstances, but they also seemed to lead to an integration of the country's strong existing resources in organic and biochemistry as well as biophysics with the science of DNA as developed in Western Europe and America since 1945 (see chapter 2 and 3). Insofar, the institutional and geographical aspects subliminally present in this book may provide stepping stones for further inquiries into the post-1970s development of the life sciences, investigating an integration of fields previously disconnected (bioenergetics, enzymology, molecular genetics, structural biology), or reciprocally a dispersion and possible transformation of the methods and the explanatory thrust of molecular biology.

Zooming in from global to local, it appears exciting to further explore two important sites at which this story unfolded—San Francisco and Munich. As different as the two urban spaces may seem at first glance—the baroque Bavarian capital in sight of the Alps, seat of a university and a major hub of the MPG, versus the quintessentially open and dynamic Californian port, a symbol of counterculture at the time as it is now for microelectronics and biotech—one may draw a parallel between them in the context of the late twentieth-century sciences and technology.

Epitomized not only in the concrete blocks of the MPI of Biochemistry, Munich became a city of progress in science and technology of 1970s West Germany (though one may think of counterculture as well). In fact, what had never been an industrial site comparable to, e.g., Berlin before, turned into the country's leading high tech city after around 1970, harboring major microelectronics and other R&D intensive businesses, as well as a large number of non-university research establishments (many from the MPG, but not exclusively).[513] A contemporary visitor to the site where Dieter Oesterhelt tinkered with the purple bugs and membrane solutions now finds not an institute or a campus, but an entire city of science and biotechnologies sprawling around the MPI at Martinsried. The unfolding of what has been baptized "Gene Valley" began with the opening of the Genzentrum and the adoption of recombinant DNA—the biocomputing project described in chapter 4 was one of the smaller endeavors when compared with the biomedical projects carried out in cooperation between academic partners and large pharmaceutical firms, or the foundations of new businesses that followed suit after 1990. As a result, what had been a mere farmers' village at the outset of this story has turned into Europe's second largest biotech site after Cambridge, UK (the third main scene of this story).[514] In this respect, Munich as a site of science and technology in the 1970s and 1980s indeed displays some parallels to the city that harbors UCSF's Parnassus Campus, where Walther Stoeckenius' group embarked on the consequential membrane project in 1969. UCSF, among a few other institutions of the Bay Area at the time, turned from an institution carrying out postwar basic research into a prime site invested in applying the new molecular biology. Soon after, it became eponymous with biotech when recombinant DNA researchers such as Herbert Boyer founded companies and turned the Bay Area, including Stanford and other universities, into the new technology's West Coast hub.[515] Probing Eric Vettel's assumption of biotechology's "countercultural origin" in San Francisco's urban space, one may ask what ferments these two cities have provided for such developments and what other ideas and projects could be unearthed (from demands of "accountable" research to alternative technologies or reorientations of existing industries, as discussed in chapter 4).

With respect to membrane science and other fields of research that were not as tightly connected to biotech as was recombinant DNA, but became linked to it later, it may furthermore be worth investigating how research and researchers have fared in the shadow of changes of institutional priorities and funding structures, of academic roles, projects, and work routines. In short, one may ask what science became under the spell of biotech, and what this spell was originally. From the 1974 UCSF press conference on a novel molecule presented as a panacea to civilizational challenges such as energy, water, or food shortages, to the articles in *Omni*-like tech-zines a decade later, to mind-boggling scenarios of bio-devices at the end of the Cold War, or to a biochip appearing in the business reports of a German chemical company, the history of molecular machinery, which could also be framed as a rather traditional investigation of molecules, appears to be invested with newness, applicability, and ever accelerating expectations of progress. How did the environment of a tech city, one may ask, influence what it meant to do science in the late twentieth century, and how did the power and promise of the biological for technology reshape important aspects of scientists' identities?

These questions bring me to an even larger riddle at the center of this book: The flip side of asking how molecules have developed into machines could also involve asking how the actors of the present story could be adequately described. Put simply: What type of scientific persona, to adapt a concept from Lorraine Daston, did the likes of Richard Henderson and Dieter Oesterhelt represent? Could one argue that this book's narrative can be read not only as the rise of a new research object, but also as that of another particular collective identity of researchers in the

life sciences in the late twentieth century, characterized by specific aspira-
tions and a specific way of life within and beyond science, as exemplified
by the plethora of rhodopsinists and membranologists that I have come
across in the course of my inquiries in Britain, West Germany, or the
United States?[516]

Strikingly, these figures as I understand them do not match with the
scientific personae known from the history of postwar molecular biology: They were not intellectuals providing philosophical or historical accounts on life or biology, such as Erwin Schrödinger or François Jacob,
and not charismatic "cult figures" such as Max Delbrück. Neither were
they openly politicized to a degree comparable to John D. Bernal, Jacques
Monod, or Linus Pauling, and they can also not be described well as
unconventional-hedonistic individuals in the style of a James D. Watson.[517]
This might be easy to explain since all of the mentioned, essentially modern figures were part of an earlier generation, studying before the war or
in its immediate aftermath, before the practice of science had significantly
changed through the expansion of research and universities. Maybe more
surprisingly, these book's actors also do not match with the persona of
the "scientific entrepreneur" as explored by Paul Rabinow at the example
of the inventor of PCR, Kary Mullis, and further characterized with cases
from biotech by Steven Shapin.[518] How to describe the predominantly male
group of professional biochemists and biophysicists encountered in the
1970s and 1980s life sciences, most of which did not display academic
disdain for commerce, but who also were not truly at home in the world
of business, and most of which (no offense) would not be well characterized by the "charismatic authority" encountered in entrepreneurial science? For some of them, biotech may have been a collateral benefit, a way
to do applied science without becoming an entrepreneur themselves, or
only a bandwagon promising attention and/or funding. Most of those encountered in this story stayed faithful to small-scale science in academic
institutions, nobody turned into a public intellectual dabbling in philosophy or politics, and the degree of self-historicization in this field has been
negligible if compared to molecular biology, recombinant DNA, or the
Human Genome Project.[519] In brief: Membrane research was neither part
of the "eighth day of creation," to pick up Horace Judson's title on early
molecular biology, nor was it a ninth.[520] A Khorana was no Watson, and
a Henderson was no Craig Venter, although their scientific credentials
certainly did not differ, and I would go as far as to stipulate that in the research described here, there were no Watsons or Venters. Might there be
a connection to the fact that these scientists did not inquire life so much
as they inquired matter, seemingly less of a philosophically or politically

charged project? In a way, this puzzle, even void of what an academic scientist of the 1970s and 1980s was beyond the quintessentially modern molecular biologist and the quintessentially postmodern biotechnologist, could be seen as an extension of the argument I have provided for biotech, where quite a number of actors fitted neither well into the category of the entrepreneurial scientist, nor into that of a countercultural activist or any other of the boxes that appear so self-evident from today's perspective.[521]

It might be an appropriate moment to pursue these questions as, since I began researching for this book in 2009, the generation described above has retired and one may suspect that their mode of research and their self-understandings have become as much a part of a near past as has the late-modern concrete architecture of the MPI at Martinsried, or UCSF's Parnassus campus.[522] Still breathing a spirit of novelty and progress, for an onlooker of a later generation, these buildings have simultaneously become monuments of a bygone time, of the moment around 1970 at which Michel Foucault and François Jacob famously diagnosed that "life," as a category and historical product of the nineteenth century, was progressively fading from the view of science. Thus, when this history of membranes and molecular machines has illustrated the tight interconnections between 1970s' and contemporary research, at the same time it brings to the fore the distance between this past and the present. And if it was the transforming force to take cells, organisms, and molecules apart and to remake their materiality that has pushed our picture of biology and ultimately ourselves far from where it was almost half a century ago, what do we make today of the once visionary, but now historical statement that life is not questioned anymore?

Abbreviations

ATP: adenosine-triphosphate
BR: bacteriorhodopsin
EM: electron microscopy
LB: Langmuir-Blodgett
ME: molecular electronics
MED: Molecular Electronic Devices
MPI: Max Planck Institute
MPG: Max Planck Society
LMB: Laboratory of Molecular Biology
NMR: nuclear magnetic resonance
PCR: polymerase chain reaction
R&D: research and development
UCSF: University of California at San Francisco

Glossary

This short glossary will help familiarize readers with technical terms of the molecular life sciences as well as designations of disciplines or research fields used frequently throughout the text. It provides a guide for understanding this story rather than general, citable definitions.

Allostery: Designates the regulation of **enzyme** activity by substances *other* than the enzyme's substrate, which bind to specific sites of an enzyme, changing its conformation and thus affecting catalysis. Allostery contributes to the regulation of metabolism (such as through feedback inhibition) and thereby cellular homeostasis.

ATP (adenosine-triphosphate): This organic molecule, a **nucleotide** carrying three phosphate groups, is the central and universal "energy currency" of cells. Burning of foodstuffs and respiration lead to the formation of ATP by phosphorylation, most of which is catalyzed by a membrane protein called **ATP-synthase** found, e.g., in **mitochondria**. Cleavage of energy-rich phosphate bonds drives all sorts of metabolic processes, from transport across membranes to syntheses of biomolecules, perception and signal transduction, or muscle movements.

ATP-synthase (also FoF1-ATPase): **Protein** complex composed of two subunits that is found in the **membranes of mitochondria** and microbes and uses the proton gradient across the membrane for the formation of **ATP**. Research on structure and molecular mechanism of ATP-synthase was central to **bioenergetics** as the protein is a centerpiece of how cells generate energy, e.g., by **oxidative phosphorylation** or photosynthesis. As the formation of ATP is nowadays explained through mechanical movements of the protein subunits, which have even been visualized, the ATP-synthase has become a poster child of a molecular machine ("rotating device").

Bacteriorhodopsin: Membrane protein belonging to the **rhodopsin** family, found in *Halobacteria*, forming purple colored patches in their cell membranes. Due to the photoreactions of BR's **co-factor** retinal, the protein transports protons across the membrane ("proton pump"). As this allows the cells to generate energy from light, BR accomplishes a simple form of photosynthesis.

Biochemistry: Since the beginning of the twentieth century, biochemistry's research program has been, in very broad terms, to explain the structure and dynamics of cells and organisms in terms of chemistry, i.e., by characterizing biological substances and their reactions. One mainstay of biochemistry has been the study of **enzymes**. The discipline's relationship to postwar **molecular biology** is complex—generally, biochemistry is older, it has been influenced strongly by nineteenth-century physiological and organic chemistry as well as by medical research, and it has focused more on understanding metabolism, whereas molecular biology was inspired strongly by physics and focused more on genetics. However, **structural biology** of biomolecules straddles both.

Bioenergetics: In the postwar life sciences, this term became used to designate research into how cells generate, transform, and utilize biological energy on the molecular level (such as through the generation of **ATP**). Bioenergetic processes are involved in, e.g., movement, metabolism, or growth. As a loosely connected, transitory research field, bioenergetics encompassed, e.g., research on membranes, **mitochondria** (cellular "power plants"), or photosynthesis, and drew on methods from physiology, biochemistry, and biophysics.

Biophysics: A research field characterized by the use of concepts and/or methods from physics to explain biological phenomena. Biophysics has ranged from the application of mathematics to biology to the use of physical instruments (e.g., in EM, optical or magnetic **spectroscopies**) to study the molecular make-up and dynamics of life. With temporally and geographically shifting meanings throughout the twentieth century, biophysics had less disciplinary coherence than **biochemistry** or **molecular biology**, and large overlaps with both.

Carotenoids: Organic molecules, responsible for many yellow or red colorations of plant materials (e.g., carrots). For their photoreactivity, organisms use carotenoids as pigments, to harvest energy or to detect light (such as through a carotenoid called retinal, the **co-factor** of **rhodopsins**, which changes its conformation upon illumination). Chemically, retinal is a derivative of vitamin A, thus linking nutrition to physiology.

Co-factor, co-enzyme: A co-factor is a small molecule attached to certain **enzymes**, on which the latter's activity depends. Carbon-based molecules attached to enzymes, such as vitamins, are called **co-enzymes**. One example is the retinal co-factor/co-enzyme of **rhodopsin** proteins. **Co-factors** are often inorganic substances such as metal ions.

Conformational change: Widely used umbrella term designating reversible changes of a molecule's spatial structure (conformation). Protein conformational changes are often induced by external stimuli (e.g., chemical reactions, binding of substrates, absorption of light) and effect their biological function. Many molecular mechanisms are explained as a series of conformational changes, i.e., as a series of movements within one protein or an interaction of several proteins.

Enzymes: Biomolecules, most of them proteins, facilitating (catalyzing) chemical reactions within cells of all organisms. Most metabolic processes, such as digestion, respiration, fermentation, build-up of body material, etc., are explained as chemical reactions carried out by specific enzymes. These latter, found in the cytoplasm or in cell membranes, are coded in the genome. As large molecules based on **peptide** chains with specific sequences (secondary structures), enzymes acquire complex and varying spatial or tertiary structures, comprising active sites, to which substrates of chemical reactions bind.

Langmuir-Blodgett (LB)-film: One or more ordered molecular layers, formed from amphiphilic organic substances such as **lipids** by deposition on a solid support; named after American physicist Irving Langmuir and technician Katherine Blodgett. LB-films have been used as models to study chemical properties and effects of monolayers and discussed as active surface or membrane technologies, e.g., in nanotechnology.

Lipids, lipid bilayer, liposome: Collective chemical term for fats and fat-like substances. Lipid molecules are amphiphilic, i.e., they are composed of a water-loving (hydrophilic) head and a water-averse (hydrophobic) tail; hence, they are often depicted as tadpole-like structures. In addition to proteins and nucleic acids, lipids are another relevant building block of all biological cells. Assembled into a double or **bilayer** (with the heads of both layers facing outside and tails facing each other), lipids are one component of **biological membranes**. In watery solutions, lipids can self-assemble into spherical structures called vesicles or **liposomes**. These latter resemble the microscopic structure of biological cells, and have been used as test tube models of them. They form the "casing" of protocells in today's synthetic biology.

Halobacterium: Genus of halophilic (salt-loving) microbes living, e.g., in salt works, the Dead Sea, or San Francisco Bay. The protein **bacteriorhodopsin** has been discovered in the species *Halobacterium salinarum*.

Membrane, biological: Bilayer composed of two monomolecular lipid films with proteins integrated into it (fluid mosaic model). Membranes are a central characteristic of all extant cellular life, forming a semipermeable barrier between organisms and their environment. Insofar, they are constitutive of cells. Moreover, membranes are the site of numerous important biological processes, from signal generation and transduction (e.g., action potentials in nerve) to the generation or transformation of energy (respiration, photosynthesis) to the transport of substances in or out of cells.

Mitochondrion: Organelle found in all cells of higher organisms (eukaryotes), surrounded by two membrane double layers. Mitochondrial membranes, harboring, e.g., the **ATP-synthase** protein, are sites of energy generation (formation of **ATP**) from respiration, i.e., the burning of carbohydrates that consumes oxygen and produces CO_2. Therefore, mitochondria are often called cellular "power plants."

Molecular biology: Influential and programmatic mid-twentieth-century research endeavor (mainly in the US, France, and UK) attempting to describe biological processes in terms of physics and chemistry. In extent historiography, molecular biology comprised molecular genetics (DNA, virus research, and microbial genetics) plus **structural biology**. To what extent molecular biology was a discipline, a vision, a research program, or an assemblage of technologies (e.g., ultracentrifuge and electron microscope) remains debated, as much as molecular biology's relationships to **biochemistry** or **biophysics**.

Nucleotide: Biological molecules composed of an organic base (purine or pyrimidine), a sugar molecule (ribose or deoxyribose), and at least one phosphate residue. The four nucleotides adenine, cytosine, guanine, and thymine are the building blocks of nucleic acids such as DNA. Moreover, nucleotides serve as **coenyzmes** of biochemical reactions, and the adenosine nucleotide carrying three phosphate residues, **ATP**, is the cell's central energy metabolite or "currency".

Oxidative phosphorylation: In this terminal part of respiratory metabolism, located in the inner mitochondrial membranes, carbohydrates and oxygen react to CO_2 and water. The chemical energy of this reaction is converted first into a proton gradient across the inner mitochondrial membrane, and subsequently into **ATP** (a process called chemiosmosis). The mechanism of oxidative phosphorylation was the subject of a controversy in **bioenergetics** of the 1960s and 1970s.

Peptide: See protein.

Polymerase chain reaction (PCR): Method to amplify genetic material in the test tube using a DNA template, a copying enzyme (DNA polymerase), and DNA building blocks (**nucleotides**). The process is carried out by specific instruments. Since PCR was invented in the 1980s, it has become a widespread method in the molecular life sciences, biotechnologies, and forensics. The process also permits researchers to introduce mutations by specifically modifying the amplified DNA, which is then reintroduced into cells (site-directed mutagenesis).

Protein: Proteins are biological macromolecules found in all cells, representing crucial components of life comparable to DNA. All proteins consist of one or more chains of amino acids linked through so-called peptide bonds (peptides are shorter chains of amino acid chains, e.g., insulin and other hormones; the term "polypeptide," designating longer chains, is synonymous with protein). The sequence of the 20 protein-forming amino acids composing a specific polypeptide is determined by the DNA coding for the respective protein. Cellular protein synthesis is carried out by the ribosome. Proteins have two central functions within cells: 1) as building materials (e.g., the collagen protein of skin and nails) and 2) as active **enzymes**, carrying out biochemical processes. A huge variety of proteins differing in sequence, size, reactivities, or cellular location (cytoplasm or membranes) is known.

Purple membrane: Specific microscopic region and functional part of the cell membrane of the microbe *Halobacterium halobium*, harboring the **bacteriorhodopsin** protein in a crystalline arrangement, generating energy from light. The purple membrane was isolated biochemically as a strikingly colored fraction by centrifugation of cell components in the late 1960s.

Recombinant DNA: Set of methods to manipulate genes by "molecular tools," such as enzymes splicing or copying DNA, in order to create organisms with altered genetic properties. The term has also been used to designate the respective field of research that has developed out of molecular biology since the early 1970s.

Recombinant DNA, some routines of which are called "molecular" or "gene cloning," is the basis of modern biotechnologies ("genetic engineering").

Retinal: See carotenoids.

Rhodopsins: Family of light-sensitive membrane proteins found both in the disc membranes of macrobes' retinae (also called "visual purple," due to its color) as well as in the cell membranes of microbes. All rhodopsins share a conserved architecture formed by 7 alpha-helices (rod-like peptide structures) spanning the membrane. Photosensitivity is due to rhodopsin's **retinal co-factor.** Upon illumination, retinal changes its conformation, starting a sequence of **conformational changes** within the protein. Organisms use rhodopsins to perceive light (photoreceptor, transmitting a signal to the nervous system) or to generate energy (as in the case of **bacteriorhodopsin).**

Spectroscopies: Set of physical methods to study the properties of cells or biological molecules (and matter generally) by exposing them to electromagnetic radiation and recording effects such as reflections or absorptions. In the late twentieth-century life sciences, both optical spectroscopies (e.g., fluorescence) and magnetic resonance spectroscopies (e.g., nuclear magnetic resonance [NMR]) were used to study the composition and structure of biological molecules, as well as the dynamics of biological processes by recording signals over time.

Vitamin A: See carotenoids.

Structural biology: Designates the part of molecular biology scrutinizing the structure of biological macromolecules. In the postwar period, structural biology's most important results were produced by X-ray crystallography, a method adapted from chemistry that allowed molecular biologists to reconstruct 3D-models of molecular structure when crystals of the respective biomolecules were exposed to X-rays. A well-known early result of structural biology was the double helix model of DNA (1953). After 1970, other methods were added to the repertoire of structural biology, such as **spectroscopies** or electron microscopic techniques reaching down to the level of molecules (e.g., cryo-EM). Structural biology, as an ensemble of methods used for a specific purpose, forms a part of **biochemistry, biophysics,** and **molecular biology.**

Notes

PREFACE

1. I should say that I encountered two fortunate exceptions, Laura Otis and Max Stadler, at the MPI for the History of Science in 2009; see Otis 1999, Stadler 2010.

2. See Introduction for references.

INTRODUCTION

3. Alberts 1994.

4. Proton pump inhibitors (as much as other molecular drugs) have been found to cause a number of serious adverse effects, which have been related to patients' conditions, over-prescription, or secondary effects of drug action (such as de-creased calcium uptake from food, leading to increases of bone fractures). For a recent metareview, see Abramowitz et al. 2016; for repercussions in the daily press, see, e.g., Span 2016. Central scientific concepts used throughout this book are briefly ex-plained in a glossary at the end of this book.

5. Kay 1993, 16ff.

6. Goodsell 2009, 10. The analogy to nanotechnology and the idea of molecular devices in this field will be discussed in chapter 4 and the Conclusion.

7. Goodsell 2009, 3.

8. de Chadarevian 2002. For an ethnographic analysis of contemporary X-ray crystallography, see Myers 2015.

9. Francoeur and Segal 2004.

10. The book's first edition was mostly illustrated with hand-drawn graphics rather than with the computer images found in the second edition in 2009.

11. Goodsell 2009, 5–6. For a discussion of the differences between the macro- and the microworld of machines, see Conclusion.

12. de Chadarevian 2002; see also Creager 2002, Creager and Morgan 2008 on the structure of viruses, or Francoeur and Segal (2004) on molecular models more generally.

13. Keller 2009a, 22. See Conclusion for a detailed discussion.

14. Goodsell 2009, 141.

15. Abi-Rached and Rose 2010, Rose 2009, 11ff.

16. Rose 2009, 26.

17. Kühlbrandt 2000.

18. See Lanyi 2004, 666. Asp and Cys refer to the amino acids aspartate and cysteine, respectively, with the numbers indicating their position in the protein's amino acid chain. Molecular distances are given in Ångström (a tenth of a nanometer), pi-electrons refer to a chemical bond, and retinal to a specific, photosensitive part of the rhodopsin protein, which is involved in light sensing (see next chapter and glossary).

19. Lipan and Wong 2006.

20. Riskin 2016, 44ff., 132ff., 1551ff.

21. Canguilhem 2008, Kant 1983 [1793], § 64; for an historical overview of this problem, Toepfer 2011.

22. Bechtel 2006, Darden and Craver 2013, Weber 2005.

23. Canguilhem 2008, 90, stated that organisms had a "greater latitude of action" than machines, including "improvisation," "utilization of occurrences," as well as pathological states. Darden and Craver 2013, 15: "A machine is a contrivance, with pre-existing, organized and interconnected parts."

24. Alberts et al. 1994, 195. See also note 26.

25. Historian and philosopher of biology Michel Morange has put this succinctly as follows: "Proteins and macromolecular protein complexes are machines; mechanisms within these machines are formed by the rigid parts of these proteins, elements of secondary structures organized to form motifs. The mechanistic vision of protein functions (and macromolecules in general) constitutes a full part of present-day biology" (Morange 2008, 36 f.).

26. The subtitle of Bruce Alberts' 1998 paper ("preparing the next generation of molecular biologists") reveals the strategic use of this concept to promote molecular biology—similar could be said about the biomedical scenarios found, e.g., in Goodsell's book. In personal communication, Michel Morange has pointed out that not all mentions of the word "machine" in connection with protein function fulfill the criteria for a mechanistic explanation: Some cases may remain unproductive metaphors, or simply follow the successful discourse of others. Different meanings as well as the development of the machine concept will be discussed in the Conclusion. On nanotechnologies, see chapter 4.

27. Grote 2014, Kohler 1973.

28. Brandt 2004, ch. 2, Kay 2000, ch. 1.

29. Branden and Tooze 1999.

30. "If you can spray them, then they are real"; Hacking 1983, 22.

31. On organismic and mechanical background metaphors (*Hintergrund-metaphorik*) in the history of philosophy and the sciences, see Blumenberg (2013, 91ff.).

32. Reynolds 2007.

33. Block 1997, 217; Tanford and Reynolds 2001, 217.

34. Keller 2016, 7.

35. Kirschner, Gerhart, and Mitchison 2000.

36. Myers 2015.

37. See note 22 for mechanism advocates; Nicholson (2013) for a critic of mechanisms. See Conclusion for a discussion of molecular mechanisms in light of contemporary research.

38. On the genetic code and informational metaphors, Brandt 2004, Kay 2000; on the turn toward biotech, Rheinberger 1995, on molecular mechanisms, de Chadarevian 2002, 272ff.. Creager and Morgan 2008 have followed up this story for Rosalind Franklin and Aron Klug's structural biology of viruses, which directly connects to the development of a novel approach to EM at Cambridge's LMB in the 1970s, as described in chapter 2. On EM, Rasmussen 1997. For an overview, Morange 2008, Olby 1990.

39. On biotechnology, see, e.g., Bud 1994, Rasmussen 2014, Yi 2014; on sequencing and genomics García-Sancho 2012; on collecting and comparing in the recent molecular life sciences, see Strasser and de Chadarevian 2011, Strasser 2012; on epigenetics and development Jablonka and Lamb 2006; on the fate of reductionism and the move to higher organisms in molecular biology, Morange 1998, 2008.

40. Rheinberger 1997, ch. 10 and following; see also Grote and Keuck 2015.

41. Klein and Lefèvre 2009, 304, Daston 2000, 2.

42. Klein and Reinhardt 2013, ix/x.

43. The quote is from Daston 2000, 2. Cf. analysis of the development of the concept of therapeutic substances, or *Wirkstoffe* (i.e., enzymes, hormones, and vitamins) by historian of science and medicine Heiko Stoff (2012).

44. Keller 2009a, 2011, 2016. On the general physical framework of active matter, Ramaswamy 2010. I thank Karin Krauthausen (Cluster of Excellence "Matters of Activity," HU Berlin) for discussions on this topic.

45. Bensaude-Vincent 2016.

46. Schäffner 2017, Tibbits 2017.

47. Creager 2017. My thanks goes also to Jennifer Rampling, apparently the prime ferment for this reactive terminology.

48. On the history of biochemistry, see e.g., Fruton 1999, Holmes 1992a/b, Kohler 1973; on the relationship of biochemistry and molecular biology, Kay 1993, de Chadarevian and Gaudillière 1996. Notably, also Garland Allen's foundational "Life Sciences in the Twentieth Century" dwells extensively on the "chemical foundations of life," especially on cell physiological and spectroscopic techniques, many consequential developments of which are discussed in chapter 2 (Allen 1978).

49. Bechtel 2006, 236ff., for a programmatic source, see Monod 1971; on structural biology of the ribosome, Rheinberger 2015.

50. Kornberg 1989, ch. 7; Rabinow 1996. In fact, DNA polymerases nowadays serve as copying agents for DNA in the omnipresent polymerase chain reaction (PCR) process; that is, they have become commercialized molecular machines forming a centerpiece of thermocyclers, i.e., the macroscopic machines that do PCR.

51. Stadler 2010; for elements toward a cultural history of membranes and surfaces, Grote and Stadler 2015; on channels in neurobiology, Trumpler 1997; on microscopic technique and membrane structure models, Liu 2018; on photosynthesis and biochemical mechanisms, Nickelsen 2015; on bioenergetics, e.g., Allchin 1996, 1997, Prebble 2013, Prebble and Weber 2003, Weber 2004.

52. Olby 1990.

53. Analyses were carried out with the ISI Web of Knowledge database, which contains electronic versions of indexed articles of many English language science journals as well as some non-English publications (e.g., German, Russian; monographs or conference abstracts are not covered, but these play a negligible role for the area and the time under analysis). Keyword searches were carried out for the respective terms as contained in the articles' "topic" data, i.e., title plus abstract. Use of the term "pump" in conjunction with "membrane" increased from 4 records in 1970 (0.05%) to 66 (0.8%) in 1990 to 568 (5.5%) in 2000 (percentage values indicate the share of the respective year's publications in the overall sum of publications detected from 1965 to 2010); the development is similar for the term "channel" in conjunction with "membrane." The increase in frequency of these terms' usage as counted by publications was higher than a baseline provided by the overall increase of publication number during the respective period (>100-fold [pump]/>500-fold [channel] versus >4-fold increase of publications from 1970 to 2000).

54. Analyses carried out as specified in note 53. The number of papers containing "protein" and "molecular mechanism" in their topic section increases from 2 in 1976 (0.008%; first record) to 123 (0.485%) in 1990 and 4040 (15.9%) in the year 2000 (percentage values indicate the share of the respective year's publications in the overall sum of indexed publications on the topic detected from 1965 to 2000. Again, the increase is higher than that of the overall number of publications indexed in ISI for the respective period (>60-fold versus c. 1.5-fold in the period from 1976 to 1990). For the term "molecular mechanism" as used in ISI-indexed publications from the areas of biochemistry/molecular biology, biophysics, cell biology, immunology, and pharmacy/pharmacology, increase is also detectable (>10-fold increase of term frequency from 1976 to 1990 as compared to >2-fold increase of overall publication number). Analyses carried out as specified in note 53.

55. See Grote and O'Malley 2011. First usage in Oesterhelt and Hess 1973 (here in the body text only). The term mechanism as found in the "topic"-data increases from 1 count in 1977 to 6 counts in 1990, 77 in 1995 and 68 in 2000, out of a total of 577 counts comprising these two terms in their "topic"-data from 1970–2000. Analyses carried out as specified in note 53. A full text search through papers indexed in JSTOR confirms this trend.

56. Boyer 1997, Deisenhofer and Michel 1989; on Boyer, see Prebble 2013.

57. Analyses carried out as described in notes 53/54. First usage of the term "molecular machine" detected in 1990 (2 counts, 0.06%), 1995 (12 counts, 0.4%) 2000 (50 counts, 1.58%), 2010 (222, 7.03%; percentage values indicate the share of the respective year's publications in the overall sum of indexed publications on the topic). Again, the rate of increase is higher than that of overall publication activity in the period (c. 100-fold as compared to c. 2-fold from 1990 to 2010). On the history of nanotechnology, Mody (2011, 2017).

58. In addition, a few published life history interviews, such as carried out by the Science History Institute, Philadelphia, have also been used. On the use of interviews in the history of science, de Chadarevian 1997; on the history and methodology of interviewing practices, te Heesen 2013; on the analysis of notebooks, Holmes, Renn, and Rheinberger 2003.

CHAPTER ONE

59. Maynard-Smith and Szathmáry 1995.

60. See Höxtermann 2000; for an overview of membrane and cell boundary research since the eighteenth century; Lombard 2014.

61. Grote 2010, Stadler 2010; the quote by Donnan is from Grote and Stadler 2015, 313.

62. That is, as molecular genetics plus structural biology, Olby 1990.

63. Delbrück 1968, 36.

64. On Delbrück's personality and role, see e.g. Kay 1993, 255 f.; on cybernetic discourse in molecular genetics, Kay 2000.

65. Lehninger 1970.

66. Pardee 1968, 632.

67. Mitochondria are intracellular organelles of eukaryotic cells accomplishing energy production (sometimes called the cell's "power plants"); chloroplasts are organelles of plants cells involved in photosynthesis.

68. As quoted in Allchin, 1997, 81. On Mitchell's analogy to fuel cells, see Grote 2010; on bioenergetics and the controversy on oxidative phosphorylation, see Prebble 2013, Prebble and Weber 2002; on intermediary metabolism biochemistry, see Holmes 1992a/b; on photosynthesis, see Nickelsen 2015.

69. Teorell 1967, 817. Quotation marks used in original.

70. Strick 2012.

71. On colloid chemistry, see Ede 2007 and Morgan 1990; on the "world of neglected dimensions," see Olby 1986.

72. The surface area determination was carried out experimentally in a "Langmuir-Adam apparatus" device, connected to the name of American physicist Irving Langmuir, consisting of a water-filled trough to which was attached a mobile, floating barrier defining part of the surface area, and a balance measuring the force exerted on the barrier. Recording the surface pressure exerted by a lipid film on this balance allowed the area covered by the film when it was spread out as a monomolecular layer to be determined. It is no slight to the illustrative, almost iconic character of this work that its result (the lipid bilayer model) was later explained as resulting from a fortunate coincidence of quantitative errors;

see Gorter and Grendel 1925. For a critique of later views on Gorter and Grendel's work, and a comparison with other studies of the time, see Lombard 2014.

73. See Liu 2018.

74. Danielli, whose scientific career started in the realm of 1930s' colloid chemistry among, for example, Sir Frederick Donnan, Neil Adam, and William Bate Hardy, carried out studies of "monofilms," layers of chemical substances, in Langmuir-Blodgett troughs. Stadler (2010, 64ff.) understands Davson and Danielli's membrane model as resulting from the transfer of this expertise in surface chemistry to the study of cell membranes.

75. Olby 1986, 299ff.; on Needham, see Teich 1973, on Schmidt, see Liu 2018.

76. Bechtel 2006, Rasmussen 1997. Keller (2002, 215ff.) mentions microscopic techniques such as phase contrast that were also used to enhance resolution, but the effects of which have escaped historical analysis.

77. Lombard 2014, Robinson 1997, 147.

78. Rasmussen 1997, 102ff.

79. Sjöstrand 1963; for discussion, see also Wolstenholme and O'Connor 1966.

80. Bernal quote from the "General discussion," in Wolstenholme and O'Connor 1966, 474. Ciba Foundation symposia are a good indicator for topics deemed important in the history of the molecular biological and biomedical sciences of the postwar era; see, for example, the symposium on *Man and His Future* with Joshua Lederberg's talk on eugenics and euphenics.

81. Letter G. Palade to K. Porter, August 30, 1954, as quoted by Rasmussen (1997, 140–41):

> I still believe that the Swedes [i.e., Sjöstrand's group, M. G.] have an advantage in resolution . . . but their ability to integrate their results in general cytology and in physiology remains nil. This is our main asset and it will be wonderful if, in addition, we could regain the lead in high resolution. Fly therefore to Berlin [i.e., to the group of Ernst Ruska, developing high resolution EM, M. G.], don't get mixed up with the Russians, and bring back an "Übermikroskop" that will show even sodium ions crossing through the membranes.

82. Davson and Danielli 1952, xi–xii. Quote is taken from the authors' preface to the first (1943) edition of the book. Davson and Danielli attribute the statement to Jacques Loeb, the godfather of a biology interested in mechanisms.

83. Reinhardt 2017.

84. Jardetzky 1966, 969–70.

85. The short paper has been quoted more than 560 times ever since, with many contemporary citations (as analyzed by ISI Web of Knowledge, May 2016). The main reason for this is that the general principle of "alternating access" to a binding site in the protein has become the dominant model for membrane transport since the late 1970s.

86. See, e.g., Davson and Danielli 1952, ch. 16; the influence of narcotics on membranes has been studied since the work of Ernest Overton in the late nineteenth century; see Lombard 2014.

87. Davson and Danielli 1952, 1.

88. Robinson 1997, 26ff.

89. A difference to carbon compounds used as tracers was that sodium and potassium ions were not assimilated into cellular materials, but merely circulated through cells and tissues. On radioisotopes in the life sciences, Creager 2013; on sodium and potassium isotopes in physiology, see Robinson (1997, 54, 89ff.) and Stadler (2010, 282ff).

90. Kennedy 1966, 434, writes about the obstacles of transferring knowledge from intermediary metabolism (referred to here as "biosynthesis") to membranes:

> In his paper (1941), [Fritz] Lipmann also considered the energetics of active transport across membranes and pointed out that here also the coupling of such an endergonic process to the utilization of ATP is to be expected. Twenty-five years later, however, in sharp contrast to the progress in our understanding of biosynthesis, very little has been learned about the biochemical basis of the specific transport of substances across cell membranes in general or about active transport in particular. Why has this problem, recognized as a central one in cell biology, proved so refractory? In part, at least, the answer may lie in the difficulty of applying chemical methods to a problem in which cellular topology plays such a large part [. . .] The methodology which has proved so successful in studying problems of biosynthesis can be summarized in the maxim: extract and purify. This approach unfortunately necessarily involves disruption of cellular architecture and the disappearance of that distinction between cytoplasm and the surrounding medium which is the heart of the transport problem.

91. On nerve, Stadler 2010; on muscle, another case in point, D. Needham 1971; also Robinson 1997.

92. Trumpler 1997, Stadler 2010.

93. Olby 1990; on biochemistry, Holmes 1992a/b.

94. The symposia at Cold Spring Harbor can be considered a nucleus of emerging molecular biologies in America. Whereas a number of meetings have been held on biochemical and biophysical topics such as this under the directorship of biophysicist Rudolf Fricke, after geneticist Milislav Demerec led the laboratory, the scope turned toward molecular genetics. Cold Spring Harbor's biophysical heritage seems by far less well-known.

95. Steinbach 1940, 244, 249. The first mention of the term "pump" seems to have been contested, with the American physiologist Robert Dean also speculating about the existence of "some sort of pump, probably located in the fibre membrane" of frog muscles, which kept the internal sodium concentration lower than the outside. Cited in Robinson 1997, 39.

96. See the quotes in Robinson 1997, 37ff. It is not easy to follow the development of the concept of the pump in Robinson's account, since he does not always differentiate between the "pump" as an actor's term, and his own contemporary use—which only goes to show how naturalized the concept is today.

97. Dean 1941, 333.

98. See figure 8.10 in Robinson 1997.

99. Robinson 1997, 126ff.

100. Skou shared the 1997 Nobel prize in chemistry (with John E. Walker and Paul D. Boyer) for his work "for the first discovery of an ion-transporting enzyme, Na+, K+ -ATPase," https://www.nobelprize.org/nobel_prizes/chemistry/laure ates/1997/; see also Robinson 1997, 127. A comparable materialization effected by a conjuncture of physiological and biochemical experimentation had taken place in muscle research already around 1940, when several researchers associated the mechanical movement of the muscle proteins (which could be seen with the bare eye, or visualized microscopically) with biochemical reactions; see D. Needham 1971.

101. Robinson 1997. Compare also the statement from 1967 that "a plethora of attempts to explain the mechanism of those substance pumps exist, but all still have hypothetical character." Schlögl 1967, 757, my translation.

102. Bergman et al. 1969, 100.

103. Delbrück 1970a. The labels "neuroscience" (F. O. Schmitt), and "neurobiology," comprise among others nerve physiology à la Hodgkin and Huxley, but also the sensory physiology and biochemistry of vision. See, for example, the Cold Spring Harbor Symposium (1965) on "Sensory physiology" (discussed below), also Stadler 2010, 308f.

104. How far the analogy to phage's role in molecular biology holds beyond the programmatic aspect as announced by Delbrück himself seems debatable— see below; Fischer 1988, 189ff.

105. Thus, a detailed review on *Phycomyces* by Bergman et al. (1969, 136) states:

> The receptor pigment of Phycomyces is not known. Studies to find it have been based on the action spectra, since no other assay to recognize the pigment is known as yet. It has been suggested that the receptor pigment might be β-carotene, retinal attached to a protein, or a flavoprotein. These compounds exhibit absorption spectra resembling the action spectra.

> On Delbrück's 1970s' research, see also next chapter.

106. The name of the Long Island research (note 94) station has become almost synonymous with molecular biology since 1945, for the symposia, the annual phage course directed by Delbrück, and Barbara McClintock's work. Interestingly enough, Cold Spring Harbor's first director in 1928 had been Rudolf Fricke, an expert in bio-electric studies of membranes, and there had already been two meetings on membrane-related issues (Surface Phenomena, 1933, and Permeability, 1940) before geneticist Milislav Demerec became head of the laboratory in 1941. On Fricke: Stadler 2010, 145ff. On cybernetic imaging in brain research, Hagner 2006, 209ff.

107. Hubbard, Bownds, and Yoshizawa 1965, 313-14; on the early receptor potential, see Cone 1965.

108. Grundfest 1965, 1.

109. Structural biology from the 1930s to the 1960s targeted not only proteins, but also viruses and, obviously, DNA. Its ramifications with the broader history of enzymology and its main accomplishments are briefly summarized as follows: In the wake of the purifications, crystallizations, and molecular weight determi-

nations of enzymatically active proteins such as urease, the concept of enzymes as catalytically active protein macromolecules has been largely accepted since the 1930s; in the postwar period, first X-ray structures of DNA, viruses, and proteins were established (Creager 2002, Creager and Morgan 2008, de Chadarevian 2002). Sequencing of various proteins from different organisms allowed correlating structure and function of these molecules in a way that has been compared to comparative anatomy (Strasser and de Chadarevian 2011). Moreover, the folding of proteins, that is, the unique spatial conformation these molecules took in solution, became related to their chemical activity through the work of Christian Anfinsen (Creager 2008). These studies, however, focused on relatively small water-soluble proteins—thus, discrete molecular species in the test tube—and not on membranes or other larger complexes and substructures of the cell.

110. See de Chadarevian, 2002, 148ff. Jim Watson's *Molecular Biology of the Gene* took a similar direction for molecular genetics.

111. On sequencing and Sanger, García-Sancho 2012.

112. Dickerson and Geis 1969, 1.

113. Dickerson and Geis 1969, 1–2.

114. Kamminga 2003.

115. Such as the cooperativity of oxygen binding, known as the "Bohr effect."

116. de Chadarevian 2002, 277.

117. Dickerson and Geis 1969, 64.

118. Pullman and Pullman 1963; Szent-Györgyi 1957.

119. de Chadarevian 2002, 272ff.

120. Blow, Birktoft, and Hartley 1969.

121. In fact, this interaction of the three amino acids—serine, histidine, and aspartic acid—to achieve an enzymatic reaction is still taught to biochemistry undergraduates today. The fact that such molecular models of enzyme action rose to prominence in the 1960s is, however, not to say that efforts to relate enzymatic function to protein structure had not been around before. American biochemist Christian Anfinsen, for example, had related amino acids in the model enzyme ribonuclease to its molecular structure since the 1950s. He achieved insights into protein structure by indirect means, such as an unfolding and refolding of the enzyme, or cleavage of parts. Thereby, specific residues could be assigned to, for example, substrate binding or catalysis in the "active center" of ribonuclease. See Creager 2008.

122. Rheinberger 1997.

123. Dickerson and Geis 1969, 109–12.

124. Lynen also compared the fatty acid synthetase to the "assembly workshops of industry," as "in both cases, the parts or components supplied from outside are fitted together and transformed piece by piece, and leave the production site only in the form of the finished product" (Lynen 1964, 132). The Ciba Symposium 1965 also focused on supramolecular organization (see above). On Lynen and biochemistry in West Germany, see Conclusion, and Deichmann 2002; for the "Riesenpartikel," Interview of author with D. Oesterhelt, MPI of Biochemistry, Martinsried, 22 January, 2009, 2.

125. Pardee 1968.

126. Interview of author with Richard Henderson, Laboratory of Molecular Biology, Cambridge, UK, April 1st, 2010, 1.

127. Sinding (2006, 940) locates this turning point for receptors slightly earlier, and in fact, first steps to isolate and characterize hormone receptors, for example, by radioactive substrates, had already taken place in the 1960s (Prüll, Maehle, and Halliwell 2009, Trumpler 1997); on the development of the gene, see Müller-Wille and Rheinberger 2012.

CHAPTER TWO

128. On channel proteins, Trumpler 1997, on receptors Prüll, Maehle, and Halliwell 2009. A shorter version of the BR story can also be found in Grote 2013a.

129. See Introduction for data on rise of publication.

130. Rasmussen 1997, 103ff. The Rockefeller Institute was renamed Rockefeller University in 1965.

131. Stoeckenius 1994. On myelin figures in 1930s/1940s membrane research, see J. Needham 1936, Liu 2018, 234f.

132. See chapter 1; Rasmussen 1997, 124ff.

133. Grote and O'Malley 2011.

134. See chapter 1.

135. Frederick Seitz papers RU RG 304.2 Series 5 General Admin Box 42 Folder 14, "Grants Walther Stoeckenius," Rockefeller Archives.

136. Stoeckenius and Rowen 1967.

137. Wolf H. Kunau in telephone conversation with M. G., October 28, 2011.

138. Stoeckenius and Kunau 1968, 344.

139. The goal of the project at the time was to prepare intracellular structures (gas vacuoles) and to check their surface structures, which are now known to comprise no membranes at all. Rheinberger (1997, 28) defines epistemic things as material entities or processes that constitute "objects of inquiry."

140. Stoeckenius 1994, Interview of author with Dieter Oesterhelt, MPI of Biochemistry, Martinsried, Jan. 22nd, 2009.

141. "CVRI's 25th Anniversary." 1984. Isidore R. Edelman, the associate director of the CVRI in the 1960s, stated regarding the CVRI Moffitt that here, clinical cardiovascular research was "leavened with the yeast of basic science." On biotech at UCSF, most notably Herbert Boyer's group, see Vettel 2006, 66ff.; on the life sciences in the Bay area, Yi 2015.

142. Interview of author with D. Oesterhelt, MPI of Biochemistry, Martinsried, 22 January, 2009, 2.

143. Letter from Oesterhelt to DFG, 15 April, 1969. File "DFG-Projekte bis ca. 1977," Oesterhelt papers, MPG Archives, AMPG III. Abt. ZA 211, Nr. 5, 94.

144. Blaurock (1982) and Stoeckenius (1994) published retrospective accounts; Oesterhelt gave a published lecture (see Stiftung Werner-von-Siemens-Ring, 2001) and has been interviewed by the author.

145. The notebook was first made available to M. G. during a visit to the MPI of Biochemistry in 2009. This and other laboratory documents will be transferred to the archives of the MPG, Berlin. The notebook is quoted hereafter as "Oe SF";

"V1"-"V60" are the numberings of the experiments as found in the book. Notebooks by Blaurock and Stoeckenius were not available at the time of research.

146. On the impact of cell fractionation regimes in molecular biology, see Creager 2002 and Rheinberger 1997.

147. Oe SF, V1–V7.

148. Oe SF, V5, 2 October, 1969, "[D]ann zentrifugieren des Dialysates wie üblich 20000 rpm SS34 30min. Ergebnis: purple pellet, red supernatant."

149. Oe SF, V6, 23 October, 1969. "hellrot," "purpurrot," "hat keine erkennbare Struktur," "Proteinwolken," "große Fladen, neben sehr klein [sic] granula."

150. Oe SF, V8, 14 November, 1969:

> Zum Purple Pigment: Aceton gibt sofortige irreversible Entfärbung; gelbliches "Protein" fällt dann mit TCE aus, Überstand ist farblos. Harnstoff bringt keine Farbveränderung, ebensowenig anschließend Zugabe von Dithionit.

151. On smell in chemistry, see the literature quoted by Reinhardt 2014.

152. Rose 2006, 11; Klein and Lefèvre 2007.

153. Interview of author with Dieter Oesterhelt, MPI of Biochemistry, Martinsried, Jan. 22nd, 2009.

154. Blaurock 1982.

155. Interview of author with Dieter Oesterhelt, MPI of Biochemistry, Martinsried, Jan. 22nd, 2009; for the prize speech, see Stiftung Werner-von-Siemens-Ring (2001, 44).

156. Oe SF, V41, 30 June, 1969.

157. Oe SF, V34, 25 May, 1970; V35A, 1 June, 1970; V38, 13 June, 1970.

158. The strategy to detect retinal co-emerged with the use of a new detergent, which, as Stoeckenius recalled, had been recommended by Blaurock since Wilkins' group had used it to extract rhodopsin from frog retinae. Stoeckenius (1994) mentioned another analogy, namely, experiments on phototaxis on *Halobacterium* that he reportedly performed after hearing a 1969 Biophysical Society plenary lecture of Max Delbrück on a phototactic bacterium that showed a similar absorption spectrum. In fact, Delbrück was National Lecturer at a Baltimore Biophysical Society meeting; however, in February 1970, he presented on *Phycomyces* sensory physiology—a story that repeatedly interacted with the present project. See Delbrück 1970b.

159. Brown 1990.

160. Chapter 1; on structural models of signal transduction in vision, Maunsbach 2008.

161. Kremer 1997, Hoffmann 2001.

162. Morton and Pitt 1957.

163. Wald 1970.

164. Stoff 2012, 133.

165. Such as in the adoption of the Warburg apparatus to study the cell physiology of the purple membrane, see below; on the impact of physical methods on 1960s' and 1970s' chemistry, see Reinhardt 2006.

166. Oe SF, V14, 18 January, 1970. "Purple: Ein Pellet aus Z1V6 verwendet (in Wasser) Ergebnis → hexagonales Muster der Membran."

167. Glatter and Kratky 1982, Olby 1986. The technique, used since the 1920s, is sometimes also called "low" or "small angle X-ray diffraction."

168. Blaurock 1982, Email Mimi Blaurock to M. G., 25 September, 2011. Even if Blaurock, who followed a meandering career path afterwards, appears as a peripheral figure in this story, his contributions were certainly crucial to a project that made the scientific lives of Oesterhelt, Henderson, and Stoeckenius. See also Wilkins 2003.

169. Blaurock 1982.

170. Reconstituting a thin film means to materially reproduce it from its components (see the experiment of Gorter and Grendel mentioned in chapter 1); on the concept of "reconstitution" in cell and synthetic biology see chapter 3.

171. Blaurock 1982.

172. Oe SF, V26, 17 February, 1970.

173. Oe SF V18, Freeze fracturing purple, 27 January, 1970. "Die Bruchstellen der Membran: präzise 60° [sign for angle], d.h. u.U. durch hexagonale Struktur bedingt." The curious effect was later explained by the fact that the drying procedure caused shrinking of the membrane film, which then ruptured along the lines formed by the units of the crystal.

174. Blaurock 1982. The evidence is that Blaurock credited another member of the lab for having prepared the samples that first showed these patterns to him.

175. Stoeckenius suggested splitting the results, whereas Blaurock preferred one joint manuscript; see Papiere D. Oesterhelt, AMPG, III. Abt. ZA211, Nr. 9, Letter W.S. to D.Oe., 23 September, 1970. One could understand the splitting of the results into two papers as a strategy to distribute credit or to enhance output. Another consideration could have been that it was improbable to find referees skilled in both of these unrelated fields—and hence rejection was less likely if the papers were sent separately to the respective experts.

176. See, for example, the works of Mark Bretscher from Sidney Brenner's department at Cambridge's LMB; Finch 2008, 172–75. Sir John R. Maddox, *Nature*'s longtime editor, pursued a multifaceted strategy with the creation of *Nature New Biology*, not least to print more papers. *Nature New Biology* remained nevertheless bound to the editorial regime of the main journal and was discontinued two years later; Baldwin 2015, 177–79.

177. Oesterhelt and Stoeckenius 1971, 152.

178. Publikationen Dieter Oesterhelt, Papiere D. Oesterhelt, AMPG, III. Abt. ZA211, V. Referee's report to *Nature*, Stoeckenius and Oesterhelt: Bacteriorhodopsin, a rhodopsin-like protein from the purple membrane of <u>Halobacterium halobium,</u> 29 March, 1971.

179. Wald 1971, *Vision and Mansions*; Papiere D. Oesterhelt, AMPG, III. Abt. ZA211, Nr.9. See the copy of a letter by George Wald to Stoeckenius forwarded to Oesterhelt, 19 February, 1971, in which he wrote: "It is that vitamin A has never been found in a bacterium, fungus, plant or even in lower invertebrate. Its first appearances have been in the phyla that possess good eyes: the molluscs & arthropods. . . . If a bacterium contains retinal, that's a big break. That's the sticky point. Everything else is reasonable." Wald himself had studied

visual pigments of various organisms and established their relationships and special adaptations to environments such as the deep sea.

180. Papiere D. Oesterhelt, AMPG, III. Abt. ZA211, Nr. 9. See Letters Stoeckenius to Oesterhelt, 19 February, 1971, 3 April, 1971. According to Oesterhelt, he learned later that the referee had, in fact, not been Wald, but his wife Ruth Hubbard, who had worked on visual physiology at Harvard with him, and continued to do so when Wald became more of a political figure in the 1970s.

181. Blaurock and Stoeckenius 1971, 154.

182. See Holmes' detailed reconstruction of the Meselson-Stahl experiment on DNA replication from early molecular biology and his attempt to map the participants' memories on archival records; Holmes 2001, 314ff.

183. Interview of author with Dieter Oesterhelt, MPI of Biochemistry, Martinsried, Jan. 22nd, 2009; personal communication J. Lanyi, UC Irvine, to M. G., March 2012.

184. D. Oesterhelt, *Die Purpurmembran aus Halobacterium halobium, Erlangen Okt. 72*, Talk manuscript, personal collection.

185. Letters Stoeckenius to Delbrück, 12 February, 1971, 15 March, 1971; Delbrück to Stoeckenius, 3 March, 1971; Max Delbrück papers, Box 20, Folder 24. In fact, the structural data seem to have been revised shortly after, with Stoeckenius admitting to Delbrück that "he was never really happy with the model proposed in the MS and I am now sure it is wrong."

186. See Stoeckenius 1971, Part 2; "Bacterial Purple," 1971; Blaurock 1972, 538–39.

187. See chapter 1 for Delbrück's attempt at a molecular biology of membranes and perception.

188. L. Jan, Studies on rhodopsin structure, localization, turnover and function—a report for the first thesis committee meeting, p. 1; Delbrück papers Box 46, Folder 12 Lily Jan.

189. See the correspondence between them: Papiere D. Oesterhelt, AMPG, III. Abt. ZA211, Nr. 9/10.

190. Laboratory notebook "München/Tübingen" (abbreviated as "Oe Mü/Tüb") to be deposited in Papiere D. Oesterhelt; see the proposal of Oesterhelt to DFG, 21 March, 1971, Folder "Projekte DFG," Papiere D. Oesterhelt, AMPG, III. Abt. ZA211, Nr.94. On Lynen's lab, see Will 2011.

191. Oe Mü/Tüb, V75 Retinyllysin aus Purpurmembrane, 6 September, 1971. "[Kulturextrakt] in 50 ml 0.1 M Tris pH 8.0 mit 200 mg NaBH4 versetzt und 30 ml Äther zugegeben. Nach 2 min. Schütteln Äther verblasen (violette Farbe kommt teilweise zurück . . .), und 40 ml H2O + 10 ml CTAB (0.1 M) zugegeben . . . Ergebnis katastrophal; fast alles RL zersetzt, nur noch Spuren nachweisbar."

192. Interview of author with Dieter Oesterhelt, MPI of Biochemistry, Martinsried, Jan. 22nd, 2009, p.4; Oesterhelt mentions this aspect also in letters to Stoeckenius, 10/12 January, 1971, Papiere D. Oesterhelt, AMPG, III. Abt. ZA211, Nr. 9.

193. In combination with the notebooks, it looks as if the set-up of this reversible assay to study the light reactions of the purple membrane in the test tube emerged piecemeal through the exchange between the two laboratories. Letter Stoeckenius to Oesterhelt, 25 February, 1971; Papiere D. Oesterhelt, AMPG, III. Abt. ZA211, Nr. 9.

See also the experiments in the notebook München/Tübingen V17, 16 January, 1971; V44, 19 April, 1971 and the recollection by Stoeckenius (1994).

194. Letter Hess to Oesterhelt, 13 March, 1972, Folder "Dortmund," MPIBC Martinsried. This folder will be transferred to the archives of MPG, Berlin. Hess referred to a talk at the so-called Mosbacher Kolloquium, an annual meeting of the West German Gesellschaft für Biologische Chemie, which was frequented by international visitors. The 1971 Kolloquium was held under the title "The Dynamic Structure of Cell Membranes." See Tagungsarchiv der Gesellschaft für Molekularbiologie at https://www.gbm-online.de/tagungsarchiv .html; last accessed 30 January 2019.

195. Chance 1999, 32; on metabolic control, see Donaghy 2013.

196. Chance 1999, 23, 40ff.

197. Krebs and Schmid 1979.

198. Letter Oesterhelt to Hess, 23 March/2 May, 1972, Hess to Oesterhelt 27 March, 1972; Folder "Dortmund" (see note 194).

199. Oesterhelt and Hess 1973.

200. For more on vintage physiology, see Grote 2013b; on the quantum yield controversy in photosynthesis and Warburg's role, Nickelsen and Govindjee 2011.

201. "Molecular mechanism" is used for the first time with reference to the purple membrane and its protein in Oesterhelt and Hess 1973, 325; for the protein's "conformational change" in analogy to rhodopsin, Oesterhelt and Hess 1973, 323.

202. One of them was biophysicist Janos Lanyi, who picked up rhodopsin research and made important contributions to the field in the 1980s and 1990s (Grote, Engelhard, and Hegemann 2014). On NASA's Exobiology Division and Ames, see Dick and Strick 2005, 31ff.

203. "CVRI's 25th Anniversary." 1984. Email Roberto Bogomolni, University of California at Santa Cruz, to M. G., 2 February, 2010.

204. Lozier and Stoeckenius 1975.

205. See for example the following quote from Stoeckenius and Lozier 1974, 773: "The purple membrane is apparently a light energy transducer which uses a mechanism quite different from the only other known biologic energy transducers, the thylakoid membrane." The term is not used prior to Oesterhelt and Hess 1973.

206. On spectroscopy and photosynthesis research generally, see Nickelsen 2015.

207. However, Stoeckenius also remembered many photobiologists being reluctant to accept the subject in the initial phase. Neither fish nor fowl, or so one could have looked at it as well, since the purple membrane had nothing to do with plant chlorophyll on the basis of its molecular structure (Stoeckenius 1994).

208. Experiments on living cells were mentioned in the discussion on the papers, however. They were begun by Oesterhelt shortly after the collaboration with Hess and are presented in the following section.

209. On Lynen's move to Martinsried, see Will (2011, 208 ff., 231); Letter Oesterhelt to DFG, 21 March, 1971, Folder "DFG-Projekte." AMPG, III. Abt., ZA211 Papiere D. Oesterhelt, Nr.94.

210. Interview of author with Dieter Oesterhelt, MPI of Biochemistry, Martinsried, Jan. 22nd, 2009.

211. J. Engel, *Die Entstehung und Funktion des Biozentrums*, Universität Basel, 1460–2010, 7; J. Engel, *Aufbrüche in der Biologie: Molekularbiologie und Systembiologie*, https://unigeschichte.unibas.ch/fakultaeten-und-faecher/phil .nat.-fakultaet/zur-geschichte-der-phil.nat.-fakultaet/biozentrum_start.html. Last accessed August 2018. Gottfried Schatz, a former collaborator of Efraim Racker at Cornell who was doing bioenergetics on mitochondria, set up a group at the Biozentrum.

212. Interview of author with Dieter Oesterhelt, MPI of Biochemistry, Martinsried, Jan. 22nd, 2009. There is no independent confirmation of this episode; however, there are good arguments as to why the transfer of knowledge from bioenergetic membrane research could have taken place in this way. The Basel researcher, Dieter Walz, was deeply involved in membrane matters at the time, including studies in transport or photosynthesis, and he had published on the very matter that he was now proposing to Oesterhelt—to measure light-induced pH-changes directly in chloroplast preparations, or in his case, cell solutions (Walz, Schuldiner, and Avron 1971).

213. On the history of bioenergetics around 1970, Grote 2010, Morange 2007, on Mitchell, Prebble and Weber 2003.

214. Prebble and Weber 2003, 91ff.

215. J. Nunn 1976, *Molecular biology and biochemistry: A third level course*. S322 Oxidative phosphorylation: program 2. VHS cassette, The Open University, c. 7 min. On the video and bioenergetics, see also chapter 3.

216. Letter Oesterhelt to DFG, 21 March, 1971, Folder "DFG-Projekte."

217. Interview of author with Hartmut Michel, MPI of Biophysics, Frankfurt a.M., 14.12.2012, 1. The head of the biochemistry department at the Munich Medical School at the time was Theodor Bücher, a student of cell physiologist Otto Warburg, and also the renowned bioenergicist Martin Klingenberg worked there at the time (Klingenberg 2005).

218. Folder "Protonenversuche/Bleichungen 1972/1973," MPIBC Martinsried. Folder will be included in Papiere Oesterhelt, AMPG, III. Abt., ZA211.

219. For "PV1" (Protonenversuch 1), Oesterhelt used "K56" (presumably "*Kultur 56*"), and as K55 and K60 can be dated to July 1972, this month emerges as a likely starting point. See Folders "Protonenversuche/Bleichungen 1972/1973," "München/Tübingen," MPIBC Martinsried. Folder will be included in Papiere Oesterhelt, AMPG, III. Abt., ZA211. See also Letters Oesterhelt to Stoeckenius, 11 July, 1972, Papiere D. Oesterhelt, AMPG, III. Abt. ZA211, Nr.9.

220. Control experiments of cells that did not contain the purple membrane were carried out as well, which did not produce the effects.

221. Interview of author with Dieter Oesterhelt, MPI of Biochemistry, Martinsried, Jan. 22nd, 2009, (transcript p. 5). Oesterhelt gave a verbatim quote of the Bavarian professor's answer: "Herr Oesterhelt, das glaube ich nicht. Ich glaube es nicht und nicht, aber eins: Ich wünsche Ihnen, dass Sie recht haben." ("Mr. Oesterhelt, I don't believe this, I believe it not and never, but one thing: I wish that you were right.")

222. Oesterhelt 1972, 1555:

> Es erscheint auf Grund dieser Versuche möglich, daß die Purpur-
> membran in der Zelle einen Protonentransport bewirkt, welcher
> die gleiche Richtung wie in Chloroplasten besitzt und in der Lage
> ist, Lichtenergie zum Aufbau eines pH-Gradienten zwischen
> Zellinnerem und Zelläußeren auszunutzen. Ob dieser Gradient
> der Zelle zur ATP-Synthese oder zum Salztransport dient, oder
> beides koppelt, ist Gegenstand weiterer Untersuchungen.

223. Letter Stoeckenius to Oesterhelt 13 October, 1972. Papiere D. Oesterhelt, AMPG, III. Abt. ZA211, Nr.9. Shortly before, the purple membrane's function as a light receptor rather than a pump was still discussed and addressed by
phototaxis experiments.

224. See, e.g., Stoeckenius 1994; Email R. Bogomolni, University of California at Santa Cruz, to M. G., 2 February, 2010. Although Bogomolni was involved in these experiments, he was not a coauthor, nor mentioned in the following publication by Oesterhelt and Stoeckenius (1973). For correspondence, see
Papiere D. Oesterhelt, AMPG, III. Abt. ZA211, Nr.9.

225. Oesterhelt and Stoeckenius 1973, 2857. If detecting the effect was easy in
principle, the cellular response contained several phases and so the interpretation
of how the different proton fluxes (ejection, influx, or both) could be explained
was not clear. Oesterhelt's initial hypothesis of proton import by the purple membrane as in the chloroplasts, presented at Erlangen, was dropped for proton export; that is, exactly the opposite phenomenon. It is interesting to note that the
hypothesis of proton export was rapidly accepted, although both effects could be
discerned in the experimental data. Their relative contribution to the overall effect depended on various factors of the environment and the cells. The origin of
the initial alkalinization, or "pH-overshoot," which Oesterhelt had seen upon illumination, and then probably stopped the measurements too early to observe the
subsequent acidification, has been controversial ever since. For review, see Henderson 1977. The debate continued in the 1980s (see, e.g., Helgerson and Stoeckenius 1985) and became silent without having reached consensus in the 1990s.

226. Oesterhelt and Stoeckenius 1973.

227. Delbrück to Oesterhelt, 24 September, 1973, Delbrück papers Box 17,
Folder 2. Presumably, Delbrück referred to the fact that *Phycomyces* membranes
contained several proteins and it was difficult to identify a material correlate of
a photoreceptor.

228. Drachev et al. 1974, 321. Thinking in terms of electrochemistry, an
electrical effect of BR was of course already implicit in the argument of Oesterhelt and Stoeckenius' 1973 paper, as the pH gradient built up upon illumination
by protons flowing out of the cell was correlated to an electrical membrane potential. The study by Drachev et al. differs in that the authors measured the electrical component of the potential directly and that they represented their findings
in terms of electric circuitry.

229. Trumpler 1997.

230. For a more detailed summary of electrophysiological studies on rhodopsins, see Grote, Engelhard, and Hegemann 2014.

231. In addition to the influential work of Aleksandr I. Oparin on the origin of life, the biochemical studies of muscle proteins by Vladimir Engelgardt could be mentioned. In the 1960s, Engelgardt headed the Biology Division and a Molecular Biology Institute of the USSR Academy of Sciences; see Levina and Sedov 2000, 427 f.; also Graham 1993; D. Needham 1971.

232. As several sources suggest, Ovchinnikov acted as a mediator securing large state funds at a time when recombinant DNA's economic and technological potentials became visible in the US. Joshua Lederberg reported after a personal meeting with Ovchinnikov in 1985 that "he had gotten Brezhnev's personal backing to modernize Soviet biology through molecular genetics [fairly explicitly to get over the Lysenko blight], for its indispensable value to medicine and agriculture." Recombinant DNA became also instrumental to the covert Soviet bioweapons program devised since these years, with which Ovchinnikov has also been associated (Leitenberg and Zilinskas 2012, 752, and references therein).

233. See chapter 1; papers by Mark Bretscher (1971, 1972) from Cambridge indicate similar directions for other proteins.

234. Elkana (1970) spoke of "concepts in flux" for the early stages of a research project.

235. de Chadarevian 2002, Hargittai 2002; see also Introduction.

236. Interview of author with Richard Henderson, Laboratory of Molecular Biology, Cambridge, UK, April 1st, 2010, p. 1.

237. Thus Hodgkin and Huxley in a 1952 publication on sodium conductance in the squid giant axon, quoted after Trumpler 1997, 62 f.

238. Prüll, Maehle, and Halliwell 2009, 150ff.

239. Interview of author with Richard Henderson, Laboratory of Molecular Biology, Cambridge, UK, April 1st, 2010, 1–2.

240. Letters R.H. to W.S., 12 October, 1972, 8 January, 1973, 19 October, 1973, 20 August, 1974, 25 March,1975, W.S. to R.H. 5 October, 10 December, 1973. Richard Henderson papers, personal collection of Richard Henderson, Cambridge, UK.

241. Interview of author with Richard Henderson, Laboratory of Molecular Biology, Cambridge, UK, April 1st, 2010, 5. This account is basically confirmed by the correspondence between the two in Henderson's papers; as well as in the acknowledgement of Blaurock (1975). On Franklin's structural work on DNA and virus structure, see Creager and Morgan 2008.

242. Creager 2002, Creager and Morgan 2008.

243. Email R. H. to M. G., 17 October, 2011. In 1953, a Nobel Prize was awarded to Frits Zernike for developing the phase-contrast method in optical microscopy.

244. On phase-contrast EM, see Huxley and Klug 1971, Unwin (1971); generally on EM, Rasmussen 1997.

245. Finch 2008, Hargittai 2002.

246. Henderson and Unwin 1975.

247. Interview of author with Richard Henderson, Laboratory of Molecular Biology, Cambridge, UK, April 1st, 2010, p. 6.

248. Unwin and Henderson 1975, plate 3.

249. Klug 1983.

250. de Chadarevian 2002.

251. Personal communication R. H. to M. G., Feb. 22nd, 2010; see also Finch 2008 (256f., 273f.).

252. Klug 1983, 575.

253. Unwin and Henderson 1975.

254. See chapter 1; on Ostwald, Olby 1986, on the fluid mosaic, Singer and Nicolson 1972, Morange 2013.

255. Henderson and Unwin 1975.

256. Henderson and Unwin, 1975, 31.

257. Singer and Nicolson 1972.

258. See Introduction, also Robinson 1997.

259. See section "From color change to molecular mechanism—optical spectroscopy"; for a later perspective, Stoeckenius, Lozier, and Bogomolni (1979).

260. Strictly speaking, viruses or the ribosome are assemblies of several macromolecules. On such models, see de Chadarevian (2002), Creager and Morgan 2008; on the BR model below, Stoeckenius 1976; personal communication Richard Henderson.

261. For review, see Grote, Engelhard, and Hegemann 2014.

262. See E. Zeitler, Recent developments in the department of electron microscopy, in: Tätigkeitsberichte Fritz-Haber-Institut 1985, 1989, AMPG, IX. Abt. Rep. 5; letters and reports from Richard Henderson's papers also describe the state of this collaborative project around 1990.

263. Steinhauser et al. 2011, 204ff.; conversation with Fritz Zemlin, group leader in the FHI's EM department, Berlin, Aug. 16th, 2012.

264. Kühlbrandt 2014.

265. The Royal Swedish Academy of Sciences, Scientific Background on the Nobel Prize in Chemistry 2017. The Development of Cryo-electron Microscopy, Oct. 4th, 2017. https://www.nobelprize.org/nobel_prizes/chemistry/laureates/2017/advanced-chemistryprize2017.pdf.

Henderson's Nobel Lecture, published while this book was in its final stages, provides more detail on his biography and work (Henderson 2018).

266. Interview M. G. with Hartmut Michel, 14.12.2012, 6ff.; on the X-ray structure of the photosynthetic reaction center, Deisenhofer and Michel 1989. With respect to the vagaries of obtaining protein crystals, Myers (2015, 61ff.) describes the patience required, including idiosyncratic, ritual-like interventions of scientists in order to make solutions crystallize.

267. Service and Stokstad 2017; see also the video and slides of Richard Henderson's Nobel speech at https://www.nobelprize.org/nobel_prizes/chemistry/laureates/2017/henderson-lecture.html.

268. John Cage, "Lecture on Nothing," In: Cage 1961, 113. Quoted in: Bernard Khoo, High-resolution structures by electron diffraction, Henderson papers, Folder HREM 1989–90, Correspondence.

269. "News from University of California San Francisco," Memo, UCSF Office of Public Information, 23 February, 1976; "Researchers uncover new kind of photosynthesis," NASA News, Press Release 76–30, 2 Mar 1976; "Basic discoveries result from find of new kind of photosynthesis," News from University of California San Francisco, Memo, UCSF Office of Public Information, 2 Mar

1976; all documents private collection. The format of the press conference to publicize research results gained an important and controversial role in recombinant DNA research and early biotech during the same period (Rasmussen 2014).

270. Brody 1977, "A strange bacteria's purple pigment . . . ," *New York Times* (5 Jul 1977), p. 17; G. Alexander, "Pigment of sea bacterium turns sunlight into energy," *Los Angeles Times* (3 Mar 1976), p. 1; Sullivan 1976, "Bacteria viewed as power source," *New York Times* (3 Mar 1976), p. 39. For more on BR and membrane technologies, see chapter 4.

271. In light of this publicized discovery narrative, the relationship between Oesterhelt and Stoeckenius deteriorated and they turned from collaborators to competitors. Interview of author with Dieter Oesterhelt, MPI of Biochemistry, Martinsried, Jan. 22nd, 2009, 5/6. Personal communication, J. Lanyi, Irvine.

272. I owe this expression to Natalie Jas, Paris. Other such moments of materialization were, for example, the isolation of drug receptors by radioactive or toxin labelling; however, in these cases, only minute quantities (micrograms) of a "receptor substance" could be isolated. See Prüll, Maehle, and Halliwell 2009, 150 ff., Trumpler 1997. Genetic approaches to membranes took a different route at the time by studying membrane phenomena through loss-of-function mutations—this was done for example by the groups of Jon Beckwith at Harvard, Maurice Hofnung at the Institut Pasteur, or H. Ron Kaback at UCLA.

273. Singer and Nicolson 1972; Morange 2013.

274. Several Nobel awards have rendered these findings more well-known, such as the 2003 prize for the structure and mechanism of a potassium channel to Roderick McKinnon, or the 2012 award to Robert Lefkowitz and Brian Kobilka for their work on the signal transducing seven-helix receptor family, which comprises rhodopsin and BR along with hormone and neurotransmitter receptors. For a review, see Vinothkumar and Henderson 2010.

275. See Sinding (2006), who speaks of them as "boundary objects." She locates this turning point for receptors slightly earlier; on the development of the gene, see Müller-Wille and Rheinberger (2012).

276. Kornberg 1989, p. 299.

CHAPTER THREE

277. Doch die Vorstellung vom Leben aus dem Automaten ist "natürlich Blödsinn," wie Diplom-Chemiker Thomas Dörper meint. Der Betreuer der Gen-Maschine weiß: "Reinste Chemikalien kommen in die Maschine hinein und ein rein chemisches Kunstprodukt kommt irgendwann wieder heraus. Chemisch gesehen ist das zwar DNA, aber biologisch gesehen ist das Produkt absolut tot." Ibelgaufts 1983, 33 (translation M. G.).

278. Russell 1987.

279. Jackson 2014.

280. Roosth 2017.

281. Bensaude-Vincent 2009b, Campos 2009.

282. On the impact of physical technique on the life sciences, Reinhardt 2006, Slater 2002.

283. Bangham, Hill, and Miller 1974, 5, 56. "Smectic" describes the layer-like arrangement of molecules in liquid crystalline phases, or "mesophases."

284. On the connection to prewar colloid chemistry, see Bangham, Hill, and Miller (1974, 1–2) and Stadler 2010.

285. Bangham 1995, 1081.

286. Barthes 1972, 214–216.

287. The term "ghosts" was used to designate membrane sacs of erythrocytes devoid of cytoplasm, which were used to study red blood cells. Davson and Danielli 1952.

288. Mueller et al. 1962a.

289. Bangham 1995, 1083. These thin films in fact appear black since they do not reflect light anymore.

290. Fischer 1988, 182f.

291. Mueller et al. 1962a; Mueller et al. 1962b.

292. Bangham 1995, Gregoriadis 1973.

293. Thus, an extensive 1974 review of "life synthesis" concluded that techniques developed to create "artificial cell envelopes" presently found "their major application in elucidating mechanistic principles of biochemical phenomena" (Widdus and Ault 1974, 30).

294. Grote 2010.

295. The 1974 paper by Racker and Stoeckenius had been quoted more than 500 times by 2014, with the highest annual rates in the late 1970s and early 1980s. It has been referred to in many textbooks on bioenergetics, such as White 2000, as well as in the philosophical analyses of Allchin 1996 and Weber 2002. Prebble (2013) is critical of a pivotal function of the study early on.

296. Schatz 1996, Allchin 2008.

297. This protein complex is also called F_oF_1-ATPase, see Glossary; Grote 2014, Prebble 2013.

298. Racker and Racker 1981, 270.

299. Racker had already published a book under the title *Mechanisms in Bioenergetics* in 1965, thereby referring to a more conventional chemical meaning of the term as in "enzymatic" or "regulatory mechanisms" (compare to the general expression of "reaction mechanisms" in chemistry; Racker 1965, v). *A New Look* (Racker 1976) comprises a curious mixture of autobiographical and historical episodes, parts of a textbook and tidbits of practical laboratory wisdom that epitomize Racker's nerdy humor and hands-on approach to science. Racker appears as a colorful and idiosyncratic personality with a tragic biography: Steeped in Viennese culture, such as the lectures of Karl Kraus, or his interest in painting and music (his brother became a psychoanalyst in Argentina), he escaped to the UK in 1938 and later on to the US. His Cornell lab became an important place on the map of biochemistry and bioenergetics in the 1960s and 1970s. Later on, he became entangled in a scientific scandal as a PhD student from his lab, Mark Spector, produced nonexistent data to detect the so-called Warburg effect of cancer cell metabolism, a long-standing controversial matter in biochemistry. See Racker and Racker 1981, Schatz 1996.

300. R. Pothier, "Life's Energy System Is Partially Duplicated," *Miami Herald*, Jan. 15th, 1972.

301. Nunn 1976, "Biochemistry and Molecular Biology Oxidative phosphorylation Programme 2"; Min. 15 and following.

302. Nunn 1976, "Biochemistry and Molecular Biology Oxidative phosphorylation Programme 2"; Min. 17 and following.

303. Delbrück to Racker, 6 June, 1973; Racker to Delbrück, 15 June, 1973, Delbrück papers Box 18, Folder 8.

304. Racker and Stoeckenius 1974.

305. Allchin 1996, Weber 2002; for an early analysis of the bioenergetics controversy in the spirit of the sociology of scientific knowledge, see Gilbert and Mulkay 1984.

306. Racker 1977, Schatz 1996, 339. Gottfried Schatz seems to blow the same horn when stating in Racker's obituary that the experiment had convinced the most "obdurate sceptics" to accept the chemiosmotic hypothesis. Marcel Weber (2002, 39) speaks of a "remarkable experiment," Michel Morange (2007, 1246) calls it a "wonderful confirmation of the mode of Mitchell"—however, not without pointing succinctly to the distinctions between Mitchell's and later versions of the chemiosmotic theory.

307. Allchin 1996.

308. Letter P. Mitchell to W. Stoeckenius, 11 June, 1973, Peter Mitchell Papers, University of Cambridge PR-G1017.

309. Letters P. Mitchell to W. Stoeckenius, 21 December, 1977, 1 March, 1978, W. Stoeckenius to P. Mitchell, 31 January, 1978. Peter Mitchell Papers, University of Cambridge PR-G1018. This was the core issue of the controversy between Mitchell and M. Wikström on the mechanism of cytochrome c oxidase function in the 1980s; see also Prebble and Weber 2003, Prebble 2013.

310. Interview of author with Dieter Oesterhelt, MPI of Biochemistry, Martinsried, Jan. 22nd, 2009, 9. Oesterhelt's account is confirmed by Mitchell's 1973 letter mentioned in note 308, in which he suggests Stoeckenius investigate bacteriorhodopsin's redox reaction, i.e., to look for a proper chemical reaction rather than conformational changes. The encounter may have taken place at the 1974 Ciba Foundation meeting on bioenergetics in London, although Mitchell is not listed amongst the participants.

311. Research on the ATP-synthase by Efraim Racker, Paul Boyer, Peter Mitchell, and others provides another example. See Loeve 2016, Prebble 2013.

312. See chapter 2. Other landmark events were e.g., the 2003 Nobel prize for the structure and mechanism of a potassium channel to Roderick McKinnon, or the 2012 award to Robert Lefkowitz and Brian Kobilka for their work on the signal transducing seven-helix receptor family, which comprises rhodopsins along with hormone and neurotransmitter receptors; Vinothkumar and Henderson 2010.

313. Interview of author with Dieter Oesterhelt, MPI of Biochemistry, Martinsried, Jan. 22nd, 2009, 9.

314. Nunn 1976 (see above), Min. 18ff.; on Dickerson and Geis, see plate 2 and chapter 1.

315. Alberts 1994, Holland et al. 2003.

316. On biobricks in synthetic biology, see Campos 2012, Morange 2009. Among many pieces of DNA and enzymes, the biobrick registry of the iGEM's

(Internationally genetically engineered machine) "Registry of Standard Biological Parts" lists a number of membrane proteins, such as receptors, channels, or pumps (e.g., halorhodopsin; see http://parts.igem.org/Protein_coding_sequences/Mem brane; last accessed Sept. 1st, 2018). Status entries such as "It's complicated!" may point to the difficulties of engineering these parts; yet, optogenetics has made successful use of such proteins, as described in the next chapter.

317. For a curious case from postwar origin of life research in India connecting back to Oparin's work, Grote 2011; for contemporary discussions, Adamala and Szostak 2013.

318. Hanczyc 2009.

319. Morgan 1990; Teich 1973 and Stadler 2010 have argued against the premature death of colloid chemistry, an impression which presumably arose after the controversy between the colloid versus the macromolecular concept of proteins was resolved in favor of the latter in the 1930s.

320. Foucault 1977, 157.

321. Jackson 2014, Laszlo 2006, Stoff 2012, 137ff.; on vitamin C, see also Reichstein 1985.

322. Slater 2002, Slater 2008, Stoff 2012.

323. Khorana produced defined, synthetic genetic messages (i.e., oligonucleotides of defined sequences) and utilized these as templates to which the ribosome, the cell's apparatus to synthesize specific proteins, gave defined "molecular answers." Kay 2000, Rheinberger 1997.

324. Olby 1996, Morange 1998.

325. As quoted in Grote 2015. Khorana's personal papers were not available for research.

326. Khorana 2000.

327. Bensaude-Vincent 2009b.

328. Khorana 2000.

329. On biomimetic chemistry, see Breslow 1972, Khorana 2000, 44, 253ff.; on the use of this concept with regard to nanotechnologies, Bensaude-Vincent 2009a.

330. Khorana 1960, 940.

331. "Synthetic Gene Works Well in Living Cell," 1976, 475.

332. This meant that recombinant cells copied and assembled the genes with their genetic apparatus as described below; "Synthetic Gene Works Well in Living Cell," 1976, 475.

333. Merrifield 1985.

334. Itakura and Riggs 1980, Khorana 2000, Merrifield 1985, Caruthers 2013. The relevance of Khorana's work for recombinant DNA was highlighted not least in a lawsuit against the patenting of the PCR process in the 1980s: The chemical giant DuPont argued against the patent holder of PCR, the biotech company Cetus, that this process was *in nuce* already present in one of Khorana's papers from the 1960s (Rabinow 2006).

335. Morange 1998, Yi 2015, 53ff.

336. Khorana 2000, 476.

337. Khorana 2000, 478; Letter Khorana to Henderson 11 January 1977, papers of Richard Henderson, personal collection of R.H. Cambridge, UK.

338. García-Sancho 2012.

339. Reinhardt 2006.

340. Ovchinnikov et al. 1979. Ovchinnikov was among the most important Soviet biochemists as well as a politically influential figure; see chapter 2. Khorana's sequence included minor corrections.

341. Dunn et al. 1981.

342. Specifically, they had excised the pump's DNA from the bacterial chromosome and pasted it into a plasmid, i.e., a vector for genetic engineering allowing the gene to be transferred, multiplied, and analyzed by the new DNA sequencing methods; on this so-called c(omplementary) DNA approach, see Morange 1998, 195f.

343. Sumper, Reitmeier, and Oesterhelt 1976; cf. Morange 1998, Rheinberger 1995.

344. Maniatis, Fritsch, and Sambrook 1982 and following editions.

345. However, keep in mind that copying, altering, and remaking DNA was much more laborious than it became after the introduction of PCR in the later 1980s (Rabinow 2006).

346. Ferretti et al. 1986; for a summary see Khorana 1988.

347. Nassal et al. 1987.

348. Jackson 2014.

349. Oprian et al. 1986.

350. For a summary of such experiments, see Kaback 1987. On Hacking's theme, see Introduction.

351. On automated nucleotide synthesis, see Itakura and Riggs 1980, also below; Morange (1998, 217) interprets the impact of this and related techniques of "directed mutagenesis" as the transition of molecular biology from a "science of observation" to a "science of action."

352. This assay was a version of Racker and Stoeckenius's 1974 reconstitution experiment, in which the pump molecules were incorporated into liposomes; see above.

353. See Introduction; Lanyi 2004.

354. Morange 1998.

355. See, for example, Gerwert et al. 1989; for review of these works, Grote, Engelhard, and Hegemann 2014.

356. Methods to display and differentiate proteins according to their molecular size or conformation, such as gel electrophoresis or mass spectrometry, then allowed probing the specific location of the cross-link within the molecule. Khorana 1980.

357. Reinhardt 2017.

358. See Conclusion for more on the dynamic view of proteins resulting from the use of these methods.

359. For review of contributions using these techniques, see Grote, Engelhard, and Hegemann 2014.

360. Whereas BR research began with biochemistry and biophysics, and genetics was used successfully only in the 1980s, the situation was the opposite for other membrane proteins such as sugar transporters. The transport of sugar across cell membranes, its regulation, and its absence in mutants had been studied phenomenologically (i.e., by monitoring cell physiology or biochemistry) since the 1950s in Jacques Monod's group at the Institut Pasteur, or in the group

of Harvard microbial geneticist Jon Beckwith. The genetic "factors" responsible for these physiological phenomena were mapped and located to the cell membranes; however, the proteins behind them were only isolated biochemically in the course of the 1980s or later on the basis of recombinant DNA technologies used to purify proteins (so-called tags attached to them and then used in affinity chromatographies). In the course of the 1990s, however, the respective missing dimension (i.e., either protein biochemistry or molecular genetics) was added, thus creating a more unified approach to and picture of membrane proteins. See Hofnung 1982, Shuman 2003, White 2000.

361. Morange 2011, Strasser 2012.

362. Lefkowitz 2004.

363. Interview of author with Dieter Oesterhelt, MPI of Biochemistry, Martinsried, Jan. 22nd, 2009, 2, my translation. Oesterhelt's description resonates with how scientists talk about running experimental systems in Hans-Jörg Rheinberger's work (Rheinberger 1997).

364. See Grote, Engelhard, and Hegemann 2014. Another case in point is the German chemist and 1980s recombinant DNA advocate in the chemical industry, Ernst-Ludwig Winnacker; see chapter 4.

365. On biotech kits, Rebentrost (2006); on chemical and biotech companies, Marschall 1999, Bertrams et al. 2013, and Hughes 2011.

366. Müller-Wille and Rheinberger 2012, Yi 2005.

367. Rheinberger 1995; on chemical syntheses, Stoff 2012 and above; on sequencing García-Sancho 2012. Dominic Berry (London) is undertaking a study of DNA synthesis and its commodification.

368. Campos 2009, Hanczyc 2009, Morange 2009.

369. Brandt 2015.

370. Foucault 1977, 145.

371. Bensaude-Vincent 2009a, Campos 2009, Keller 2002, Morange 2009, Roosth 2017.

372. Foucault 1977, 139. Interestingly, a broad and method-centered scientific review on the problem of "life synthesis" was published in 1974, with a preface by membrane physiologist James F. Danielli, who discussed its potentials for food supply, population growth, and the problems of industrial growth (Widdus and Ault 1974).

373. Ibelgaufts 1983, also Winnacker 1990. Possibly, Winnacker's use of the term could be related to James F. Danielli's in 1974 and/or that of molecular geneticist Jack Szybalski a few years later; on Szybalski, the label "synthetic biology" vs. that of "genetic engineering" in the 1970s, and the eclipse of the former from the late 1970s to around 2000, see Campos 2009.

CHAPTER FOUR

374. Chorost 2009, Colapinto 2015.

375. Cheng 2007.

376. On bioelectronics in medicine, see Tracey 2015.

377. The "light switches" engineered into the rat brain's motor neuron membranes, which allow researchers to selectively modify the cells' activity (i.e.,

to trigger or repress action potentials by switching on lights of different wavelengths) are actually photosensitive ion pumps and channels belonging to the rhodopsin family, which function, e.g., as photoreceptors, or "molecular eyes" of algae. Since the 1980s, they have been studied by membrane biophysicists such as Ken Forster or Peter Hegemann—the former was a PhD student of Max Delbrück, the latter of Dieter Oesterhelt. See Beck 2014; Grote, Engelhard, and Hegemann 2014; Hegemann and Sigrist 2013.

378. Chorost 2009, McAuliffe 1981 (see below).

379. Müggenburg 2014, Bud 2010.

380. McCray 2013, 13.

381. Tucker 1984. The author, Jonathan B. Tucker, later on became an expert on biological and chemical weapons (Shapiro 2011); on Conrad, see Fogel, Matsuno, and Paton 2001, Mody 2017a, 105.

382. Tucker 1984, 42; Conrad 1985.

383. An enzyme electrode biosensor is composed of a conventional semiconductor device coupled to an electrode (as known from, e.g., pH-meters) covered with an enzyme, the activity of which would catalyze a chemical reaction. Such technologies were around since the 1960s as devices to determine, for instance, blood glucose concentrations in medicine. Some protagonists of biosensors have speculated about future uses of these rather mundane devices in or on the body, e.g., as automated monitors and therapies for diabetic patients. Clarke 1987.

384. Stadler 2014.

385. Tucker 1984, 41.

386. Different meaning of the term were already distinguished in 1982 in the *New Scientist*, Yanchinski 1982.

387. The "DNA computer," invented in the 1990s on the basis of PCR and gel electrophoresis techniques, analyzes and displays strand-pairing of DNA molecules according to their sequence similarities. As a logical gate based on the combinatorial properties of DNA base pairing, the use of different DNA probes in this assay has allowed certain mathematical problems to be resolved. However, it did not comprise an interface that coupled wet test tube biochemistry to microelectronics; see Gratzer 2009, 221–230. On DNA chip technologies, such as microarrays, Yi 2010. It should be noted that the term "biocomputing" has also been used to designate the uses of (conventional) computers in biology— again, something completely different; see November 2012.

388. The term "biochip fever" was later used by a Japanese researcher for this phase; see Aizawa 1991, 107.

389. Mody 2017a, 7ff.

390. Choi and Mody 2009, Mody 2017a.

391. Mody 2017a, 92ff., 94.

392. Ceruzzi 2003, 23–24.

393. Kaminuma and Matsumoto 1991, Roland and Shiman 2002, Stadler 2014.

394. Carter 1982, iii.

395. Carter 1987; McCray 2013.

396. Drexler 1981, 5276.

397. Drexler 1990 [1986], on recombinant DNA, see below.

398. McAuliffe also was an editor of *Omni*. I owe my thanks to Cyrus Mody for drawing my attention to her work (Mody 2017a).

399. McCray 2012.

400. McAuliffe 1981, 58.

401. Gibson 1986.

402. Both Michael Conrad and cell biologist Lynn Margulis (see below) were editors of the journal *BioSystems*, for example, which published a number of speculative or generalizing articles (Fogel, Matsuno, and Paton 2001); see also the book by Stuart Hameroff (1987) discussed below.

403. David Kaiser and W. Patrick McCray (2016) have introduced the label "groovy science" for a number of projects and people in between academe and counterculture.

404. On Michel's work, chapter 2; on Kuhn and Germany, see below. For a characterization of these scientists, see Conclusion. *Biocomputers — The Next Generation from Japan*, a 1988 book alluding to the country's prior "Fifth Generation Initiative" to develop silicon-based parallel processing computers at the beginning of the decade, presented a number of exchanges between computer technology and the life sciences, such as active membranes, molecular circuit devices, and "artificial neural devices." These appear under the umbrella of a 10-year national project on bioelectronic devices funded by the Japanese Ministry of International Trade and Industry and involved companies such as NEC, Mitsubishi, and Sanyo (Kaminuma and Matsumoto 1991; Miyasaka, Koyama, and Itoh 1992; Watsuji et al. 1995).

405. Chance et al. 1980, 32. On Chance and his spectroscopic research into protein molecular mechanisms, see chapter 2. His coauthor was Paul Mueller, who had synthesized artificial membranes in the 1960s (see chapter 3). Interestingly, Fritz Lipmann, a German émigré biochemist studying biological energetics (see chapter 1 and 2) had pondered about "molecular technology" of biological structures and microelectronics already around 1970, allegedly inspired by miniaturization debates (Lipmann 1971).

406. Chance et al. 1980, 38.

407. Chance et al. 1980, 32; on the microscience issue, see Krumhansl and Pao 1979. Kaplan and Radin 2011 have analyzed a similar journal, *Chemical & Engineering News*, for a later period, and introduced the term "para-scientific" to designate that these media exist next to and in a way parasite on outlets for original research, whereas they are not part of popular science either.

408. Chance et al. 1980, 38; Drexler 1981 quotes the *Physics Today* piece in his seminal *PNAS* paper on "molecular engineering."

409. This method had already been introduced in the 1930s by General Electric physico-chemist Irving Langmuir and assistant Katherine Blodgett; see Tanford and Reynolds 2001. On the MPI of Biophysical Chemistry, see Henning and Kazemi 2016, 290ff.

410. The terms in German were "Baukasten," "Schaltelemente von molekularer Dimension," and "Informationsverarbeitung." Kuhn 1965, 258, 265; see also Kuhn's contribution to the second MED meeting (Kuhn 1987).

411. Thus British nanotechnologist Richard Jones in retrospect, quoted from Mody 2017a, 108.

412. On the "gene jockeys," see Rasmussen 2014, on the scientific entrepreneur, Shapin 2008; on Khorana and recombinant DNA in protein studies, see chapter 3.

413. Ulmer 1982, 215; McCray 2013, 167ff.

414. Koselleck 1989 used these concepts to conceive of modern historicity since the French Revolution and industrialization; however, they seem fruitful also to understand this very specific moment in the history of technology around 1980.

415. Rasmussen 2014, 190–91.

416. Tucker 1984, 47.

417. McAuliffe 1984, 20.

418. Helmreich 2000.

419. McAuliffe 1984, 20.

420. Hameroff 1987. This was an early usage of the term nanotechnology, though, NB, spelled differently. Hameroff's vision differs from Drexler's in that he aimed at using the scanning-tunnel microscope, an instrument to create ordered molecular structures, rather than relying on molecular assembly (Mody 2011).

421. Hameroff 1987, xx.

422. See e.g., Hameroff and Penrose 1996.

423. Strick 2015, 96; see also Grote 2016.

424. Sapp 2015, 116f. Even if Margulis' spirochaete theory was criticized from the beginning, and in fact was not supported by molecular evidence such as endosymbiosis, the idea to seek for "molecules of cognition" conserved in evolution was not so outlandish. An established researcher such as Eric Kandel had surmised at the 1983 Cold Spring Harbor Symposium that molecular neurobiology's next challenge was: "the possibility—indeed, the likelihood—that many molecules important for the higher nervous functions of humans may be conserved in evolution and found in the brains of much simpler animals, and, moreover, that some of these molecules may not even be unique to the cells of the brain, but may be used generally by cells throughout the body." Kandel thereby referred to the second messenger system (cAMP-system) of cell signaling, but the hope of finding general molecular principles of cognition was related. Kandel 1983, 907.

425. Margulis and Sagan 1986, 252; also referred to by Hameroff 1987, 68f.

426. The phrasing "bionics at the molecular scale" was used in a journalistic account of the German biochip project; see note 446.

427. Müggenburg (2014) draws a line from interwar approaches of biological design headed as "*Biotechnik*" to postwar American bionics, which developed in the context of cybernetics. On a comparable 1980s bionic project in the context of energy technologies, Grote 2016.

428. On *CoEvolution Quarterly*, published by Brand after the Whole Earth Catalogue, Turner 2006, 120ff.

429. Conrad 1985; on the perceived crisis of the "Boolean dream," Stadler 2014.

430. Tucker 1984, 40.

431. Conrad 1986, 60.

432. Grote 2016.

433. Bud 1993; on bionics/*Bionik*, Müggenburg 2014.

434. Tucker 1984, 47.

435. Hong 1986, 223–24; Hong 1989, 105.

436. Michael Conrad adopted a similar view in the proceedings of a 1993 meeting on ME and biocomputing, arguing that the developments in rhodopsin research "have served in the past decade to demonstrate beyond a shadow of doubt that biomolecular materials can be tamed to perform functional services. Other biomolecules may shoot to the fore with similar alacrity in the next decade." Conrad 1995, 100; see also Bräuchle, Hampp, and Oesterhelt 1991; Schick, Lawrence, and Birge 1988; and Vsevolodov 1998.

437. "Bakteriorhodopsin—Ein Biopolymer für die optische Informationsverarbeitung, Teil 2," 1989.

438. Bud 1994, Jasanoff 1985.

439. On Winnacker, see chapter 3 and note 373; on the gene centers Rebentrost 2006, 73f.

440. Bundesminister FT, 1986, 54ff.; Bundesminister FT 1989; Projektleitung Biologie, Ökologie, Energie 1989; Interview of author with Norbert Hampp, University of Marburg, Munich, 15.12.2012, Stiftung Werner-von-Siemens-Ring 2001.

441. See, e.g., Wacker Chemie 1992.

442. On Solvay, see Bertrams et al. 2013, on Monsanto and semiconductor material production Lécuyer and Brock 2006, on German chemical companies, Rebentrost 2006. Wacker Unternehmenskommunikation (2003) present aspects of Wacker's corporate history.

443. Such as slower photocycles or shifted absorption maxima to make the protein compatible with laser technologies; see Hampp et al. 1992; Oesterhelt, Soppa, and Krippahl 1987; Oesterhelt et al. 1990.

444. Oesterhelt, Bräuchle, and Hampp 1991, 427.

445. Wacker Chemie 1991, 19.

446. "Dabei können die Forscher auf Jahrmillionen höchst effizienter 'Entwicklungsarbeit' der Natur zurückgreifen." Biospeicher. Roter Bazillus 1992, 213. The original expression was "Bionik in Molekülgröße," in: M. Weiner (1993), Revolution in der Materialforschung, Philipp Morris Forschungspreis, brochure, private collection.

447. Bud 1994.

448. Lécuyer and Brock 2006.

449. Interview of author with Norbert Hampp, University of Marburg, Munich, 15.12.2012, 5, 13; Hampp, Bräuchle, and Oesterhelt 1990, 83.

450. Mody 2017a, 108; see also note 411.

451. Interview of author with Norbert Hampp, University of Marburg, Munich, 15.12.2012, 2, 5; Hampp, Bräuchle, and Oesterhelt 1990, 83.

452. Carter 1982, 221–22; for more on such fundamental issues, see also Conclusion and Jones 2004.

453. Browne 1987. Probably the article related Birge's research on rhodopsins with synthetic co-factors. On the Soviet version of the biochip, labeled as "Biochrome," see Vsevolodov et al. 1986, on military applications, Vsevolodov 1998.

454. Birge 1995, 68. Ovchinnikov had died in 1988.

455. Vsevolodov 1998, 155.

456. Interview of author with Norbert Hampp, University of Marburg, Munich 15.12.2012, 4;. Email N. Abdualev (former member of Ovchinnikov's team) to M.G., 6.12.2009; for publications from the time, see, e.g., Vsevolodov et al. 1986.

457. Bud 1994, 2010; on current biomolecular nanotechnologies, see Hampp 2006.

458. Hampp 2006.

459. One should add that this was not so straightforward in practice, as problems occurred not only regarding the amounts produced, but also with respect to, e.g., photochemical side reactions that potentially disturbed the technological use of the protein's "switching" process.

460. Tanford and Reynolds 2001, 3; Gratzer 2009, 205ff.; for a philosophical analysis of machine concepts in research on the ATP-synthase, Loeve 2016.

461. On Drexler's assemblers, Jones 2004, McCray 2012, 171; on machines, molecules, and organisms, see Introduction and Conclusion.

462. Matsuno 1991, 323–24.

463. On a similar line, it seems worthwhile to study how far concepts such as complexity, adaptability, self-organization, or evolution have transformed with the rise of theoretical biophysics in the 1970s and 1980s—think of Manfred Eigen's hypercycle or Isabelle Stengers and Ilya Prigogine's widely received theories (Prigogine and Stengers 1984); see also Conclusion.

464. Barad 2007, Keller 2011, 2016.

465. Interview of author with Norbert Hampp, University of Marburg, Munich 15.12.2012, 11.

466. Mody 2017a.

467. Robert Bud (1994) construes biotechnology exactly in this way from the beginning of the century to the 1970s; however, he narrows his perspective as well when discussing the 1980s.

468. With a perspective on probe microscopy, these different roots of nanotechnology and usage of the "nano"-label by Drexler and others are analyzed in detail in Mody 2011, 175ff.

469. Mody 2017a.

470. Müller-Wille and Rheinberger 2012; Rasmussen 2014, 190–91.

471. Jones 2004, 5.

472. The original term is "*vorsichtig*," meaning also careful. Koselleck 1989, 374.

CONCLUSION

473. Foucault 1974, 207ff.; Jacob 2002 [1970], 318, see also pp. 17, 325. Elsewhere in the book, Jacob also discussed enzymes and pumps, artificial protein synthesis, and the "fine anatomy of molecules," see p. 286, 313.

474. For a phenomenological analysis of the concept, see Soentgen 2008.

475. Latour and Woolgar 1979.

476. See Ursula Klein and Wolfgang Lefèvre's historical ontology of chemical substances in the early modern period: "In classical chemical analysis, the

most elusive—molecular chemical structure—was entwined with the most quotidian—dirt and smell" (Klein and Lefèvre 2007, 304); Daston 2000, 2 (also Introduction, note 41).

477. Shapin 1989; the role of technicians and manual labor with "vintage technique" in Oesterhelt's membrane research in analyzed in detail in Grote 2013b.

478. On viruses, Brandt 2004, Creager 2002; on enzymes, Grote 2014, Kohler 1973.

479. Keller 2011, 2016, Bensaude-Vincent 2016; see also Karen Barad's related argument for an "agential realism" (Barad 2007).

480. For these and other cases, Gratzer 2009, Jones 2004; on "molecular vitalism," Kirschner, Gerhart, and Mitchison 2000. Natasha Myers (2015) contrasts scientists' "lively renderings" of protein structure models (such as when protein movements are verbalized as "breathing," or in animations) to their "de-animated mechanistic ontology" of these molecules (p. 185). It remains to be said, however, that similar descriptions of protein behavior have equally drawn inspiration from mechanical devices (such as hinges, gates, turnstiles, etc.), both in lab discourse as well as in publications.

481. Loeve 2015, Jones 2004, Gratzer 2009.

482. Keller 2016, 7; see also Introduction.

483. Craver and Darden 2013.

484. Canguilhem 2008, Nicholson 2013, Riskin 2016; see Introduction.

485. On the cell as a "collection of molecular machines," Alberts 1998, Alberts et al. 1904, Canguilhem 2008; see also Introduction.

486. Haraway 1985, 69; see also epigraph to chapter 4. On theories of self-organization around 1980, see Adorf 2016, Keller 2009b.

487. See Introduction; Goodsell 1993 and 2009.

488. Reynolds 2007. Andrew Reynold's book on the history of cell biology, which was published while this manuscript was in its final stage, provides arguments on metaphors' turn to literality and a reshaping of entities that resonate very well with my take (Reynolds 2018).

489. The localization and conformation of dynamic protein regions is resolved very poorly or not at all in X-ray crystallographic models. In addition to such "blind spots," there are even more fundamental limitations of these models, due to the quality of the crystals (minimal wavelength at which date could be obtained) or the intrinsic reliability of what is displayed in the model (the so-called B-factors); see Branden and Tooze 1991; on graphic modeling practices, de Chadarevian 2002, Morange 2011.

490. The method could be applied to structural biology in connection with biochemical preparation methods, such as cloning, site-directed mutagenesis, overexpression, isotope labeling, or chemical modifications (see chapter 3). The first full NMR structure determinations of proteins, performed in parallel with X-ray methods in order to lend credibility to the new method, were carried out in the 1980s; Reinhardt 2017, Steinhauser 2014.

491. Such sets of a protein's different conformers can be understood in terms of chemical thermodynamics as coexisting in equilibrium with each other, i.e., the individual molecules in a population are present in the respective states with different probabilities and fluctuate between them. Enzymological studies have

often enough supported such a dynamic view long before NMR methods were around. Yet, such classical "wet" biochemical assays did not result in direct visualizations of molecular structure, possibly also because these assays were somehow low-tech, they had less impact, and only a visualizing method such as NMR has given the dynamic perspective more credibility against the dominance of models from X-ray crystallography.

492. Myers (2015) speaks of different "renderings" of the molecular world. On recent membrane protein studies combining crystallography with magnetic resonance spectroscopies as well as "wet" biochemical methods to establish detailed structure-function models, see, e.g., Celia et al. 2016, Bordignon, Grote, and Schneider 2010.

493. Jones 2004, 64ff.

494. Loeve 2015.

495. See e.g., Holmes 1992 a/b, Kohler 1973.

496. On Mitchell, see Prebble and Weber 2003.

497. Creager 2017.

498. See e.g., Borck 2005, Brain and Wise 1994, Hagner 2006, Rabinbach 1992, Stadler 2010; on optogenetics, e.g., Hegemann and Sigrist 2013.

499. Kamminga and Cunningham 1995, Nyhart 2017; on metabolism, e.g., Landecker 2013, Grote and Keuck 2015; on origin of life research, Strick 2012.

500. Kay 1993, see Introduction.

501. Abi-Rached and Rose 2010, Rose 2009.

502. Creager 2017.

503. Morange 1998; on the recent rise of interest in developmental biology and epigenetics, see Jablonka and Lamb 2006.

504. Lipan and Wong 2006; see Introduction.

505. Another important case straddling neurobiology, physiology, and pharmacology are receptors of bodily substances (e.g., hormones) or physico-chemical stimuli (light, pressure, etc.); see Lefkowitz 2004, Sinding 2006.

506. García-Sancho 2012, Stevens 2013, Strasser and de Chadarevian 2011, Strasser 2012.

507. Olby 1990, de Chadarevian 2002.

508. For the period before 1970, photosynthesis is another good case in point (Nickelsen 2015).

509. For an early account of the human genome project, Cook-Deegan 1994; see also the references in note 470.

510. Edgerton 2008; for an episode from 1970s molecular membrane research highlighting the productivity of "old" methods from organic chemistry or a "vintage instrument" such as the Warburg apparatus, see chapter 2; in more detail, Grote 2013b.

511. On the postwar situation of molecular genetics and biochemistry in West Germany, Deichmann (2001, 2002); on Martinsried and the MPG, Heßler 2007.

512. On EMBL, de Chadarevian 2002.

513. Heßler 2007, 63ff.

514. Heßler 2007, 167ff.

515. Hughes 2011, Vettel 2006, Yi 2015. Furthermore, these two cities not only stood in for the scientific and technological projections of the 1970s, but

have also been hubs of sub- and countercultures and so new ways of life in a very different sense — to connect the actors and themes of science and technology with these broader sociocultural developments would be an exciting subject of further studies.

516. On the scientific persona, Daston 2015. Both Henderson (born 1945) and Oesterhelt (born 1940) received their PhDs in the late 1960s from within the national centers of their fields at the time (Cambridge's LMB and the MPI of Cell Chemistry, respectively); both embarked on a topic that was novel and somewhat radical by methodical and conceptual standards of the early 1970s, and both built their careers on it by using their expertise of existing fields (enzymology and structural biology) to venture on something new. Oesterhelt held the post of a director at the Max Planck Institute for Biochemistry from 1979 to 2009, and his department of "Membrane Biochemistry" was an international hub for PhD students, postdocs, and visitors, bringing forth a number of internationally influential figures. Richard Henderson became director of Cambridge's LMB and received a Nobel prize in 2017.

517. On Delbrück as a "cult figure," Kay 1993, 255 f.; on Watson as a type who had overcome the ascetic ideal of the scientist, similar to physicist Richard Feynman, Shapin (2008, 217 f.). A plethora of (auto)biographical accounts of the mentioned and other molecular biologists exist, a number of which are critically discussed in Abir-Am (1991); for a more casual but still insightful account, see Judson 1996.

518. Rabinow 2006, Shapin 2008, ch. 7; on the expansion of postwar science and its consequences for the role of the scientist, Kaiser 2004, Mody 2016.

519. For example, Cook-Deegan 1994 and note 470; on Craig Venter, Brandt (2015).

520. Judson 1996.

521. See chapter 4; for another example from bionic research in Germany, Grote 2016. In an essay on the "excluded middle" of 1970s science, beyond counterculture and entrepreneurialism, Mody (2017b) has introduced the term "square scientist," for mostly white, straight, middle class men, who were no zealots, but involved in applying science to technology development or socially relevant projects and who frequently go unnoticed in historical accounts. It seems that a number of actors, especially from chapter 4, match this characterization, although the specifics of being a scientists in the US versus West Germany would need further analysis to sharpen this category.

522. Dieter Oesterhelt retired as director of the MPI in 2009, and he closed his laboratory in 2014, as did a number of other important players in this story around the same time — Henderson, Stoeckenius' former staff member Roberto Bogomolni at UC Santa Cruz, Janos Lanyi of UC Irvine, or Hartmut Michel at Frankfurt's MPI of Biophysics. Har Gobind Khorana (born 1922) and Walther Stoeckenius (born 1921), a generation older, passed away in 2011 and 2013, respectively.

Sources

Archival Material

Max Delbrück papers, California Institute of Technology Archives, Pasadena.

Fritz-Haber-Institut der MPG, Tätigkeitsberichte, Archiv der Max-Planck-Gesellschaft Berlin (AMPG), IX. Abt. Rep. 5

Peter Mitchell papers, GBR/0012/MS, University of Cambridge, Cambridge UK.

Dieter Oesterhelt papers, AMPG, III. Abt. ZA 211

Frederick Seitz papers RU RG 304.2, Rockefeller Archives, Sleepy Hollow, NY.

Personal Papers

Richard Henderson papers, private collection of Richard Henderson, Cambridge, UK. These papers will be transferred to the Archives of the MRC-LMB, Cambridge, UK.

Interviews

All interviews were carried out by the author, transcribed entirely or in parts and authorized by the interviewees. Rather than following a fixed set of questions, the topics for discussion were chosen individually for each interviewee according to the interest to this story. All interviews are unpublished and kept in private collection.

Interview with Norbert Hampp, Philipps-Universität Marburg, Munich, Dec. 15th, 2012.

Interview with Richard Henderson, Medical Research Council Laboratory of Molecular Biology, Cambridge, UK, April 1st, 2010.

Interview with Hartmut Michel, MPI of Biophysics, Frankfurt a.M., Dec. 14th, 2012.

Interview with Dieter Oesterhelt, MPI of Biochemistry, Martinsried, Jan. 22nd, 2009.

References

Abi-Rached, Joelle M., and Nikolas Rose. 2010. "The Birth of the Neuromolecular Gaze." *History of the Human Sciences* 23, no. 1: 11–36.

Abir-Am, Pnina. 1991. "Nobelesse Oblige: Lives of Molecular Biologists." *Isis* 82: 326–43.

Abramowitz, Jason, Punam Thakkar, Arton Isa, Alan Truong, Connie Park, and Richard M. Rosenfeld. 2016. "Adverse Event Reporting for Proton Pump Inhibitor Therapy." *Otolaryngology–Head and Neck Surgery* 155, no. 4: 547–54.

Adamala, Katarzyna, and Jack W. Szostak. 2013. "Nonenzymatic Template-Directed RNA Synthesis inside Model Protocells." *Science* 342, no. 6162: 1098–1100.

Adorf, Henrik. 2016. "'La Nouvelle Alliance' Chaos-Theorie und die Wiederverzauberung der Natur in den achtziger Jahren." *Nach Feierabend. Zürcher Jahrbuch für Wissensgeschichte* 12: 117–32.

Aizawa, Masuo. 1991. "Biodevice computers." In *Biocomputer: The Next Generation from Japan*, edited by Tsuguchika Kaminuma and Ken Matsumoto, 107–119. London: Chapman & Hall.

Alberts, Bruce. 1998. "The Cell as a Collection of Protein Machines: Preparing the Next Generation of Molecular Biologists." *Cell* 92: 291–94.

Alberts, Bruce, D. Bray, J. Lewis, M. Raff, K. Robert, and J. D. Watson. 1994. *Molecular Biology of the Cell.* 3rd ed. New York: Garland Science.

Alexander, George. 1976. "Pigment of Sea Bacterium Turns Sunlight into Energy." *Los Angeles Times*, Mar. 3, 1976.

Allchin, Douglas. 1996. "Cellular and Theoretical Chimeras: Piecing Together How Cells Process Energy." *Studies in History and Philosophy of Science Part A* 27, no. 1: 31–41.

Allchin, Douglas. 1997. "A Twentieth-Century Phlogiston: Constructing Error and Differentiating Domains." *Perspectives on Science* 5, no. 1: 81–127.

Allchin, Douglas. 2008. "Efraim Racker (1917–1991)." In *New Dictionary of Scientific Biography*, vol. 6, edited by Noretta Koertge. Detroit: Charles Scribner's Sons/Thomson Gale.

Allen, Garland. 1978. *Life Sciences in the Twentieth Century*. Cambridge: Cambridge University Press.

"Bacterial Purple." 1971. *Nature* 233: 238.

"Bakteriorhodopsin—Ein Biopolymer für die optische Informationsverarbeitung, Teil 2." 1989. *Werk + Wirken* 7: 3–7.

Baldwin, Melinda. 2015. *Making "Nature": The History of a Scientific Journal*. Chicago: University of Chicago Press.

Bangham, Alec D. 1995. "Surrogate Cells or Trojan Horses: The Discovery of Liposomes." *BioEssays* 17, no. 12: 1081–88.

Bangham, Alec D., M. W. Hill, and N. G. A. Miller. 1974. "Preparation and Use of Liposomes as Models of Biological Membranes." In *Methods in Membrane Biology*, edited by Edward D. Korn, 1–68. New York: Springer.

Barad, Karen. 2007. *Meeting the Universe Halfway: Quantum Physics and the Entanglement of Matter and Meaning*. Durham: Duke University Press.

Barthes, Roland. 1972. "The Structuralist Activity." In *Critical Essays*, 213–20. Evanston: Northwestern University Press.

Bechtel, William. 2006. *Discovering Cell Mechanisms. The Creation of Modern Cell Biology*. Cambridge Studies in Philosophy and Biology. Cambridge: Cambridge University Press.

Beck, Christina. 2014. "Einzeller bringen Licht in die Neurobiologie." *Max-Planck-Forschung* 3: 19–25.

Bensaude-Vincent, Bernadette. 2009a. "Biomimetic Chemistry and Synthetic Biology: A Two-way Traffic across the Borders." *HYLE: International Journal for Philosophy of Chemistry* 15, no. 1: 31–46.

Bensaude-Vincent, Bernadette. 2009b. "Synthetic Biology as a Replica of Synthetic Chemistry? Uses and Misuses of History." *Biological Theory* 4, no. 4: 314–18.

Bensaude-Vincent, Bernadette. 2016. "From Self-Organization to Self-Assembly: A New Materialism?" *History and Philosophy of the Life Sciences* 38: 1.

Bergman, Kostia, P.V. Burke, E. Cerdá-Olmedo, C.N. David, M. Delbrück, K.W. Foster, E.W. Goodell, et al. 1969. "Phycomyces." *Bacteriological Reviews* 33, no. 1 (March): 99–157.

Bertrams, K., E. Homburg, N. Coupain, G.K. Hentenryk, and P. Mioche. 2013. *Solvay: History of a Multinational Family Firm*. Cambridge: Cambridge University Press.

The Biological Laboratory, ed. 1940. *Permeability and the Nature of Cell Membranes*. Cold Spring Harbor Symposia on Quantitative Biology, vol. 8. New Bedford: Darwin Press.

"Biospeicher. Roter Bazillus." 1992. *Der Spiegel* 7: 213–14.

Birge, Robert R. 1995. "Protein-Based Computers." *Scientific American*, March, 90–95.

Blaurock, Allen E. 1972. "The Purple Eye of a Bacterium." *New Scientist* 53: 538–9.

Blaurock, Allen E. 1975. "Bacteriorhodopsin: A Trans-Membrane Pump Containing [alpha]-Helix." *Journal of Molecular Biology* 93, no. 2 (April 5): 139–58.

Blaurock, Allen E. 1982. "Analysis of Bacteriorhodopsin Structure by X-Ray Diffraction." In *Biomembranes Part I: Visual Pigments and Purple Membranes II Volume 88*, edited by Lester Packer, 124–32. New York: Academic Press.

Blaurock, Allen E., and Walther Stoeckenius. 1971. "Structure of the Purple Membrane." *Nature New Biology* 233, no. 39: 152–55.

Block, Steven M. 1997. "Real Engines of Creation." *Nature* 386: 217–19.

Blow, David M., Jens J. Birktoft, and Brian S. Hartley. 1969. "Role of a Buried Acid Group in the Mechanism of Action of Chymotrypsin." *Nature* 221: 337–40.

Blumenberg, Hans. 2013. *Paradigmen zu einer Metaphorologie*. Frankfurt am Main: Suhrkamp.

Borck, Cornelius. 2005. *Hirnströme: Eine Kulturgeschichte der Elektroenzephalographie*. Göttingen: Wallstein.

Bordignon, Enrica, Mathias Grote, and Erwin Schneider. 2010. "The Maltose ATP-binding Cassette Transporter in the 21st Century—Towards a Structural Dynamic Perspective on Its Mode of Action." *Molecular Microbiology* 77, no. 6: 1354–66.

Boyer, Paul D. 1997. "The ATP Synthase—a Splendid Molecular Machine." *Annual Review of Biochemistry* 66, no.1: 717–49.

Brain, Robert M., and M. Norton Wise. 1994. "Muscles and Engines: Indicator Diagrams and Helmholtz's Graphical Methods." In *Universalgenie Helmholtz*, edited by Lorenz Krüger, 124–45. Berlin: Akademie Verlag.

Branden, C.I., and J. Tooze. 1991. *Introduction to Protein Structure*. New York: Garland Publications.

Branden, C.I., and J. Tooze. 1999. *Introduction to Protein Structure*. 2nd ed. New York: Garland Publications.

Brandt, Christina. 2004. *Metapher und Experiment. Von der Virusforschung zum genetischen Code*. Göttingen: Wallstein.

Brandt, Christina. 2015. "Thesen zur Autorschaft in den modernen Biotechnologien. Craig Venter und die synthetische Biologie." In *Erzählung und Geltung. Wissenschaft zwischen Autorschaft und Autorität*, edited by S. Azzouni, S. Böschen, and C. Reinhardt, 259–88. Weilserwist: Velbrück Wissenschaft.

Bräuchle, Christoph, Norbert Hampp, and Dieter Oesterhelt. 1991. "Optical Applications of Bacteriorhodopsin and Its Mutated Variants." *Advanced Materials* 3, no. 9: 420–28.

Breslow, R. 1972. "Centenary Lecture: Biomimetic Chemistry." *Chemical Society Reviews* 1, no. 4: 553–80.

Bretscher, Mark S. 1971. "Human Erythrocyte Membranes: Specific Labelling of Surface Proteins." *Journal of Molecular Biology* 58, no. 3: 775–81.

Bretscher, Mark S. 1972. "Asymmetrical Lipid Bilayer Structure for Biological Membranes." *Nature New Biology* 236: 11–12.

Brody, Janet E. 1977. "A Strange Bacteria's Purple Pigment, Which Uses Light to Generate Energy, May Yield Scientific Goldmine." *New York Times*, July 5, 1977.

Brown, A.D. 1990. *Microbial Water Stress Physiology. Principles and Perspectives*. Chichester: Wiley.

Browne, Malcolm W. 1987. "Vision Chemical is Found to Absorb Radar," *New York Times*, August 18, 1987.

Bud, Robert. 1994. *The Uses of Life. A History of Biotechnology*. Cambridge: Cambridge University Press.

Bud, Robert. 2010. "From Applied Microbiology to Biotechnology: Science, Medicine and Industrial Renewal." *Notes and Records of the Royal Society* 64, suppl. 1: 17–29.

Bundesminister für Forschung und Technologie. 1986. *Angewandte Biologie und Biotechnologie. Programm der Bundesregierung 1985 – 1988*. Bonn.

Bundesminister für Forschung und Technologie. 1989. *Programmreport Biotechnologie*. Bonn.

Cage, John. 1961. *Silence*. Middletown: Wesleyan University Press.

Campos, Luis. 2009. "That Was the Synthetic Biology That Was." In *Synthetic Biology: The Technoscience and Its Societal Consequences*, edited by Markus Schmidt, Alexander Kelle, Agomoni Ganguli-Mitra, and Huib Vriend, 5–21. Dordrecht: Springer Netherlands.

Campos, Luis. 2012. "The BioBrick™ Road." *BioSocieties* 7, no. 2: 115–39.

Canguilhem, Georges. 2008. *Knowledge of Life*. New York: Fordham University Press.

Carter, Forrest L. ed. 1982. *Molecular Electronic Devices*. New York: Marcel Dekker.

Carter, Forrest L. ed. 1987. *Molecular Electronic Devices II*. New York: Marcel Dekker.

Caruthers, Marvin H. 2013. "The Chemical Synthesis of DNA/RNA: Our Gift to Science." *Journal of Biological Chemistry* 288, no. 2: 1420–27.

Celia, Hervé, Nicholas Noinaj, Stanislav D. Zakharov, Enrica Bordignon, Istvan Botos, Monica Santamaria, Travis J. Barnard, William A. Cramer, Roland Lloubes, and Susan K. Buchanan. 2016. "Structural Insight into the Role of the Ton Complex in Energy Transduction." *Nature* 538, no. 7623: 60–65.

Ceruzzi, Paul E. 2003. *A History of Modern Computing*. 2nd ed. Cambridge, MA: MIT Press.

Chance, Britton, Paul Mueller, Don De Vault, and L. Powers. 1980. "Biological membranes." *Physics Today* 33: 32–38.

Chance, Britton. 1999. Oral history interview by Sally Smith Hughes, University of Pennsylvania, Philadelphia, PA, February 3 and 5, July 7, 1999, transcript #0179 Philadelphia: Science History Institute.

Cheng, Ingfei. 2007. "The Beam of Light That Flips a Switch That Turns on the Brain." *The New York Times*, Aug. 14, 2007. http://www.nytimes.com/2007/08/14/science/14brai.html.

Choi, Hyungsub, and Cyrus C. M. Mody. 2009. "The Long History of Molecular Electronics." *Social Studies of Science* 39, no. 1: 11–50.

Chorost, Michael. 2009. "Algae and Light Help Injured Mice Walk Again." *Wired Magazine*, Oct. 19, 2009. http://www.wired.com/2009/10/mf_optigenetics/.

Clarke, Leland C. 1987. "The Enzyme Electrode." In *Biosensors: Fundamentals and Applications*, edited by Anthony P.F. Turner, Isao Karube, and George S. Wilson, 3–12. Oxford: Oxford University Press.

Colapinto, John. 2015. "Lighting the Brain. Karl Deisseroth and the Optogenetics Breakthrough." *The New Yorker*, May 18, 2015. http://www.newyorker.com/magazine/2015/05/18/lighting-the-brain.

Cone, Richard A. 1965. "The Early Receptor Potential of the Vertebrate Eye." *Cold Spring Harbor Symposia on Quantitative Biology* 30: 483–91.

Conrad, Michael. 1985. "On Design Principles for a Molecular Computer." *Communications of the ACM* 28, no. 5: 464–80.

Conrad, Michael.1986. "The Lure of Molecular Computing." *IEEE Spectrum* 23, no. 10: 55–60.

Conrad, Michael. 1995. "Introduction—Proceedings of MEBC 93." *Biosystems* 35: 99–100.

Cook-Deegan, Robert M. 1994. *The Gene Wars: Science, Politics, and the Human Genome*. New York: W. W. Norton.

Craver, Carl F., and Lindley Darden. 2013. *In Search of Mechanisms: Discoveries across the Life Sciences*. Chicago: University of Chicago Press.

Creager, Angela N. H. 2002. *The Life of a Virus: Tobacco Mosaic Virus as an Experimental Model, 1930–1965*. Chicago: University of Chicago Press.

Creager, Angela N. H. 2008. "Anfinsen, Christian B." In *The New Dictionary of Scientific Biography*, edited by Noretta Koertge, 76–82. Detroit: Charles Scribner's & Sons/Thompson Gale.

Creager, Angela N. H. 2013. *Life Atomic: A History of Radioisotopes in Science and Medicine*. Chicago: University of Chicago Press.

Creager, Angela N. H. 2017. "A Chemical Reaction to the Historiography of Biology." *Ambix* 64: 343–59.

Creager, Angela N. H., and Gregory J. Morgan. 2008. "After the Double Helix." *Isis* 99, no. 2 (June 1): 239–72.

"CVRI's 25th Anniversary." 1984. *UCSF Magazine* 7, no. 1: 2–27.

Danielli, James F. 1974. "Genetic Engineering and Life Synthesis: An Introduction to the Review by R. Widdus and C. Ault." *International Review of Cytology* 38: 1–5.

Daston, Lorraine. 2000. "Introduction: The Coming into Being of Scientific Objects." In *Biographies of Scientific Objects*, edited by L. Daston, 1–14. Chicago: University of Chicago Press.

Daston, Lorraine. 2015. "Die wissenschaftliche Persona. Arbeit und Berufung." In *Zwischen Vorderbühne und Hinterbühne: Beiträge zum Wandel der Geschlechterbeziehungen in der Wissenschaft vom 17. Jahrhundert bis zur Gegenwart*, edited by Theresa Wobbe, 109–36. Bielefeld: Transcript.

Davson, Hugh, and James Frederic Danielli. 1952. *The Permeability of Natural Membranes*. 2nd ed. Cambridge: Cambridge University Press.

Dean, Robert B. 1941. "Theories of Electrolyte Equilibrium in Muscle." *Biological Symposia* 3: 331–38.

De Chadarevian, Soraya. 1997. "Using Interviews to Write the History of Science." In *The Historiography of Contemporary Science and Technology*, edited by Thomas Söderqvist, 51-71. Amsterdam: Harwood.

De Chadarevian, Soraya. 2002. *Designs for Life: Molecular Biology after World War II.* Cambridge: Cambridge University Press.

De Chadarevian, Soraya, and Jean-Paul Gaudillière. 1996. "The Tools of the Discipline: Biochemists and Molecular Biologists." *Journal of the History of Biology* 29: 327–30.

Deichmann, Ute. 2001. *Flüchten, Mitmachen, Vergessen. Chemiker und Biochemiker in der NS-Zeit.* Weinheim: Wiley-VCH.

Deichmann, Ute. 2002. "Emigration, Isolation and the Slow Start of Molecular Biology in Germany." *Studies in History and Philosophy of Science Part C: Studies in History and Philosophy of Biological and Biomedical Sciences* 33, no. 3 (2002): 449–71.

Deisenhofer, Johann, and Hartmut Michel. 1989. "The Photosynthetic Reaction Center from Purple Bacterium *Rhodopseudomonas viridis.*" *Embo Journal* 8, no. 8: 2149–70.

Delbrück, Max. 1968. "Molecular Biology-The Next Phase." *Engineering and Science* 32, no. 2: 36–40.

Delbrück, Max. 1970a. "A Physicist's Renewed Look at Biology: Twenty Years Later." *Science* 168, no. 3937 (June): 1312–5.

Delbrück, Max. 1970b. "Phycomyces—Work in Progress." Biophysical Society Abstracts. Fourteenth Annual Meeting, Table of Contents.

Dick, Steven J., and James E. Strick. 2005. *The Living Universe: NASA and the Development of Astrobiology.* New Brunswick: Rutgers University Press.

Dickerson, Richard E. 1997. "Irving Geis, Molecular Artist, 1908–1997." *Protein Science* 6, no. 11: 2483–84.

Dickerson, Richard E., and Irving Geis. 1969. *The Structure and Action of Proteins.* Menlo Park: W. A. Benjamin.

Donaghy, Josephine. 2013. "Autonomous Mathematical Models: Constructing Theories of Metabolic Control." *History and Philosophy of the Life Sciences* 35, no. 4: 533–52.

Drachev, Lel A., A. A. Jasaitis, A. D. Kaulen, A. A. Kondrashin, E. A. Liberman, I. B. Nemecek, S. A. Ostroumov, A. Yu. Semenov, and V. P. Skulachev. 1974. "Direct Measurement of Electric Current Generation by Cytochrome Oxidase, H^+-ATPase and Bacteriorhodopsin." *Nature* 249, no. 5455: 321–24.

Drexler, K. Eric. 1981. "Molecular Engineering: An Approach to the Development of General Capabilities for Molecular Manipulation." *Proceedings of the National Academy of Sciences of the United States of America* 78, no. 9: 5275–78.

Drexler, K. Eric. 1990 [1986]. *Engines of Creation: The Coming Era of Nanotechnology.* New York: Anchor Press.

Dunn, R., J. McCoy, M. Simsek, A. Majumdar, S. H. Chang, U.L. Rajbhandary, and H. G. Khorana. 1981. "The Bacteriorhodopsin Gene." *Proceedings of the National Academy of Sciences of the United States of America* 78, no. 11: 6744–48.

Ede, Andrew. 2007. *The Rise and Decline of Colloid Science in North America, 1900–1935: The Neglected Dimension*. Aldershot: Ashgate Publishing.

Edgerton, David. 2008. *The Shock of the Old: Technology and Global History since 1900*. London: Profile Books.

Elkana, Yehuda. 1970. "Helmholtz' 'Kraft': An Illustration of Concepts in Flux." *Historical Studies in the Physical Sciences* 2: 263–98.

Engel, J. "Die Entstehung und Funktion des Biozentrums." Last accessed November 2015. https://unigeschichte.unibas.ch/fakultaeten-und-faecher/phil.nat.-fakultaet/zur-geschichte-der-phil.nat.-fakultaet/biozentrum_start.html.

Ernster, Lars, R. W. Estabrook, E. C. Slater. 1974. *Dynamics of Energy-Transducing Membranes*. Amsterdam: Elsevier.

Ferretti, L., S. S. Karnik, H. G. Khorana, M. Nassal, and D. D. Oprian. 1986. "Total Synthesis of a Gene for Bovine Rhodopsin." *Proceedings of the National Academy of Sciences of the United States of America* 83, no. 3: 599–603.

Finch, John. 2008. *A Nobel Fellow on Every Floor: A History of the Medical Research Council Laboratory of Molecular Biology*. Cambridge: Medical Research Council—Laboratory of Molecular Biology.

Fischer, Ernst-Peter. 1988. *Das Atom der Biologen: Max Delbrück und der Ursprung der Molekulargenetik*. München: Piper.

Fogel, David B., Koichiro Matsuno, and Ray Paton. 2001. "In Memoriam—Michael Conrad." *BioSystems* 61, no. 1 (2001): 3–4.

Foucault, Michel. 1974. *Die Ordnung der Dinge: Eine Archäologie der Humanwissenschaften*. Frankfurt a.M.: Suhrkamp.

Foucault, Michel. 1977. "Nietzsche, Genealogy, History." In *Language, Counter-Memory, Practice: Selected Interviews and Essays*, edited by D. F. Bouchard, 139–64. Ithaca: Cornell University Press.

Francoeur, Eric, and Jérôme Segal. 2004. "From Model Kits to Interactive Computer Graphics." In *Models: The Third Dimension of Science*, edited by S. de Chadarevian and N. Hopwood, 402–27. Stanford: Stanford University Press.

Fruton, Joseph S. 1999. *Proteins, Enzymes, Genes: The Interplay of Chemistry and Biology*. New Haven: Yale University Press.

Gamow, George, and Martinas Yčas. 1968. *Mr. Tompkins inside Himself: Adventures in the New Biology*. London: Allen and Unwin.

García-Sancho, Miguel. 2012. *Biology, Computing, and the History of Molecular Sequencing: From Proteins to DNA, 1945–2000*. London: Palgrave Macmillan.

Gerwert, K., B. Hess, J. Soppa, and D. Oesterhelt. 1989. "Role of Aspartate-96 in Proton Translocation by Bacteriorhodopsin." *Proceedings of the National Academy of Sciences of the United States of America* 86, no. 13: 4943–47.

Gibson, William. 1986. *Count Zero*. London: Victor Gollancz.

Gilbert, G. Nigel, and Michael Mulkay. 1984. *Opening Pandora's Box*. Cambridge: Cambridge University Press.

Glatter, O., and O. Kratky. 1982. *Small Angle X-ray Scattering*. Amsterdam: Academic Press.

Goodsell, David S. 1993. *The Machinery of Life*. New York: Springer.

Goodsell, David S. 2009. *The Machinery of Life*. 2nd ed. New York: Springer.

Gorter, Evert, and F. Grendel. "On Bimolecular Layers of Lipoids on the Chromocytes of the Blood." *Journal of Experimental Medicine* 41 (1925): 439–43.

Graham, Loren. 1993. *Science in Russia and the Soviet Union.* Cambridge: Cambridge University Press.

Gratzer, Walter. 2009. *Giant Molecules: From Nylon to Nanotubes.* Oxford: Oxford University Press.

Greogriadis, Gregory. 1973. "Molecular Trojan Horses." *New Scientist* 60, no. 878 (Dec. 27): 890–93.

Grote, Mathias. 2010. "Surfaces of Action: Cells and Membranes in Electrochemistry and the Life Sciences." *Studies in History and Philosophy of Science Part C: Studies in History and Philosophy of Biological and Biomedical Sciences* 41, no. 3: 183–93.

Grote, Mathias. 2011. "Jeewanu, or the 'Particles of Life.'" *Journal of Biosciences* 36, no. 4: 563–70.

Grote, Mathias. 2013a. "Purple Matter, Membranes and 'Molecular Pumps' in Rhodopsin Research (1960s–1980s)." *Journal for the History of Biology* 46: 331–68.

Grote, Mathias. 2013b. "Vintage Physiology: Otto Warburg's 'Labor-Kochbücher' und Apparaturen." *NTM Zeitschrift für Geschichte der Wissenschaften, Technik und Medizin* 21, no. 2: 171–85.

Grote, Mathias. 2014. "From Enzymes to 'Molecular Machines': Materiality in Research on Rhodopsins, 1970s." In *Objects of Chemical Inquiry*, edited by Ursula Klein and Carsten Reinhardt, 343–68. Sagamore Beach: Science History Publications.

Grote, Mathias. 2015. "Khorana, Har Gobind." In *eLS.* John Wiley & Sons. https://doi.org/10.1002/9780470015902.a0002845.

Grote, Mathias. 2016. "Das Patchwork der Mikroben: Bio-Technologien jenseits der großen Erzählungen." *Nach Feierabend. Zürcher Jahrbuch für Wissensgeschichte* 12: 15–34.

Grote, Mathias, Martin Engelhard, and Peter Hegemann. 2014. "Of Ion Pumps, Sensors and Channels—Perspectives on Microbial Rhodopsins between Science and History." *Biochimica et Biophysica Acta Bioenergetics* 1837, no. 5: 533–45.

Grote, Mathias, and Maureen O'Malley. 2011. "Enlightening the Life Sciences: The History of Halobacterial and Microbial Rhodopsin Research." *FEMS Microbiology Reviews* 35, no. 6: 1082–99.

Grote, Mathias, and Lara Keuck. 2015. "Conference Report 'Stoffwechsel: Histories of Metabolism,' workshop organized by Mathias Grote at Technische Universität Berlin, November 28–29th, 2014." *History and Philosophy of the Life Sciences* 37: 210–18.

Grote, Mathias, and Max Stadler. 2015. "Introduction: Surface Histories." *Science in Context* 28: 311–15.

Grundfest, Harry. 1965. "Electrophysiology and Pharmacology of Different Components of Bioelectric Transducers." *Cold Spring Harbor Symposia on Quantitative Biology* 30: 1–14.

Hacking, Ian. 1983. *Representing and Intervening: Introductory Topics in the Philosophy of Natural Science.* Cambridge: Cambridge University Press.

Hagner, Michael. 2006. *Der Geist bei de Arbeit: Historische Untersuchungen zur Hirnforschung.* Göttingen: Wellstein.

Hameroff, Stuart R. 1987. *Ultimate Computing: Biomolecular Consciousness and NanoTechnology.* Amsterdam: North-Holland.

Hameroff, Stuart R., and Roger Penrose. 1996. "Orchestrated Reduction of Quantum Coherence in Brain Microtubules: A Model for Consciousness." *Mathematics and Computers in Simulation* 40: 453–80.

Hampp, Norbert. 2006. "Nanobiotechnology Enables New Opportunities in Material Sciences: Bacteriorhodopsin as a First Example." In *Bionanotechnology: Proteins to Nanodevices*, edited by V. Renugopalakrishnan and Randolph V. Lewis, 209–16. Dordrecht: Springer Netherlands.

Hampp, Norbert, Christoph Bräuchle, and Dieter Oesterhelt. 1990. "Bacteriorhodopsin Wildtype and Variant Aspartate-96 → Aspargine as Reversible Holographic Media." *Biophysical Journal* 58, no. 1: 83–93.

Hampp, Norbert, Andreas Popp, Dieter Oesterhelt, and Christoph Bräuchle. 1992. "Zubereitungen von Bakteriorhodopsin-Varianten mit erhöhter Speicherzeit und deren Verwendung." German patent DE000004226868A1, filed August 13th, 1992, and issued Feb. 17th, 1994.

Hanczyc, Martin M. 2009. "The Early History of Protocells: The Search for the Recipe of Life." In *Protocells: Bridging Nonliving and Living Matter*, edited by S. Rasmussen, M. A. Bedau, L. Chen, D. Deamer, D. C. Krakauer, N. H. Packard, and P. F. Stadler, 3–18. Cambridge, MA: MIT Press.

Haraway, Donna. 1985. "A Manifesto for Cyborgs: Science, Technology, and Socialist Feminism in the 1980s." *Socialist Review* 15, no. 2: 65–107.

Hargittai, Istvan. 2002. "Richard Henderson [Interview]." In *Candid Science II: Conversations with Famous Biomedical Scientists*, edited by Istvan Hargittai and Magdolna Hargittai, 297–305. London: Imperial College Press.

te Heesen, Anke. 2013. "Naturgeschichte des Interviews." *Merkur. Deutsche Zeitschrift für Europäisches Denken* 4, no. 67: 317–28.

Hegemann, Peter, and Stephan Sigrist. 2013. *Optogenetics.* Berlin: De Gruyter.

Helgerson, S.L., and Walther Stoeckenius. 1985. "Transient Proton Inflows during Illumination of Anaerobic *Halobacterium halobium* Cells." *Archives of Biochemistry and Biophysics* 241, no. 2: 616–27.

Helmreich, Stefan. 2000. *Silicon Second Nature: Culturing Artificial Life in a Digital World.* Berkeley: University of California Press.

Henderson, Richard. 1977. "Purple Membrane from *Halobacterium halobium.*" *Annual Review of Biophysics and Bioengineering* 6: 87–109.

Henderson, Richard. 2018. "From Electron Crystallography to Single Particle CryoEM (Nobel Lecture)." *Angewandte Chemie International Edition* 57: 2–24.

Henderson, Richard, and P. Nigel T. Unwin. 1975. "Three-Dimensional Model of Purple Membrane Obtained by Electron Microscopy." *Nature* 257, no. 5521: 28–32.

Henning, Eckhart, and Marion Kazemi. 2016. *Handbuch zur Institutisgeschichte der Kaiser-Wilhelm-/Max-Planck-Gesellschaft zur Förderung der Wissenschaften 1911-2011. Daten und Quellen.* Vol 1. Berlin: MPG. https://www.archiv-berlin.mpg.de/60874/MPG_Handbuch_zur_Institutsgeschichte_Bd_2_Tb_1_E-Book.pdf.

Heßler, Martina. 2007. *Die kreative Stadt: Zur Neuerfindung eines Topos.* Bielefeld: Transcript.

Hoffmann, Christoph. 2001. "Zwei Schichten: Netzhaut und Fotografie, 1860/1890." *Fotogeschichte* 21: 21–38.

Hofnung, Maurice. 1982. "Presentation of the Maltose System and the Workshop." *Annales de l'Institut Pasteur Microbiology* 133A: 5–8.

Holland, I. Barry, Susan P. C. Cole, Karl Kuchler, and Christopher F. Higgins. 2003. *ABC Proteins: From Bacteria to Man.* London: Elsevier Science.

Holmes, Frederic L. 1992a. "Between Biology and Medicine: The Formation of Intermediary Metabolism. Four Lectures Delivered at the International Summer School in the History of Science Uppsala, July 1990." Berkeley Papers in the History of Science 14. Berkeley: University of California, Office for History of Science and Technology.

Holmes, Frederic L. 1992b. "Intermediary Metabolism, Tissue Slices and Manometers." In *The Right Tools for the Job: At Work in Twentieth-Century Life Sciences*, edited by Joan H. Fujimura and Adele H. Clarke, 151–71. Princeton: Princeton University Press.

Holmes, Frederic L. 2001. *Meselson, Stahl, and the Replication of DNA: A History of "The Most Beautiful Experiment in Biology."* New Haven: Yale University Press.

Holmes, Frederic, Jürgen Renn, and Hans-Jörg Rheinberger. 2003. *Reworking the Bench: Research Notebooks in the History of Science.* Archimedes. New Studies in the History and Philosophy of Science and Technology. Vol. 7. New York: Kluwer.

Hong, Felix T. 1986. "The Bacteriorhodopsin Model Membrane System as a Prototype Molecular Computing Element." *BioSystems* 19: 223–36.

Hong, Felix T., ed. 1989. *Molecular Electronics: Biosensors and Biocomputers.* New York: Plenum.

Höxtermann, Ekkehard. 2000. "Physiologie und Biochemie der Pflanzen." In *Geschichte der Biologie. Theorien, Methoden, Institutionen, Kurzbiographien*, edited by Ilse Jahn, 499–536. Heidelberg, Berlin: Spektrum Akademischer Verlag.

Hubbard, Ruth, Deric Bownds, and Tôru Yoshizawa. 1965. "The Chemistry of Visual Photoreception." *Cold Spring Harbor Symposia on Quantitative Biology* 30: 301–15.

Hughes, Sally S. 2011. *Genentech: The Beginnings of Biotech.* Chicago: University of Chicago Press.

Huxley, Hugh Esmor, and Aaron Klug. 1971. "New Developments in Electron Microscopy." *Philosophical Transactions of the Royal Society of London B* 261: 1–230.

Ibelgaufts, Horst. 1983. "'Spart halt Zeit.' Münchner Biochemiker produzieren mit einem Synthese-Automaten Gene per Knopfdruck." *Die Zeit* 22, May 27, 1983.

Itakura, Keiichi, and Arthur D. Riggs. 1980. "Chemical DNA Synthesis and Recombinant DNA Studies." *Science* 209, no. 4463: 1401.

Jablonka, Eva, and Marion J. Lamb. 2006. *Evolution in Four Dimensions. Genetic, Epigenetic, Behavioral, and Symbolic Variation in the History of Life.* Cambridge, MA: MIT Press.

Jackson, Catherine M. 2014. "Synthetical Experiments and Alkaloid Analogues: Liebig, Hofmann, and the Origins of Organic Synthesis." *Historical Studies in the Natural Sciences* 44, no. 4: 319.

Jacob, François. 2002 [1970]. *Die Logik des Lebenden, Eine Geschichte der Vererbung.* Frankfurt a.M.: Fischer Verlag.

Jardetzky, Oleg. 1966. "Simple Allosteric Model for Membrane Pumps." *Nature* 211, no. 5052: 969–70.

Jasanoff, Sheila. 1985. "Technological Innovation in a Corporatist State: The Case of Biotechnology in the Federal Republic of Germany." *Research Policy* 14, no.1: 23–38.

Jones, Richard A. L. 2004. *Soft Machines: Nanotechnology and Life.* Oxford: Oxford University Press.

Judson, Horace Freeland. 1996. *The Eighth Day of Creation: Makers of the Revolution in Biology.* Expanded ed. Cold Spring Harbor: Cold Spring Harbor Laboratory Press.

Kaback, H. Ronald. 1987. "Molecular Biology of Active Transport: From Membrane to Molecule to Mechanism." *Harvey Lectures* 83: 77.

Kaiser, David. 2004. "The Postwar Suburbanization of American Physics." *American Quarterly* 56, no. 4: 851–88.

Kaiser, David, and W. Patrick McCray. 2016. *Groovy Science: Knowledge, Innovation, and American Counterculture.* Chicago: University of Chicago Press.

Kaminuma, Tsuguchika, and Ken Matsumoto, eds. 1991. *Biocomputer: The Next Generation from Japan.* London: Chapman & Hall.

Kamminga, Harmke. 2003. "Biochemistry, Molecules and Macromolecules." In *Companion to the History of Science in the 20th Century,* edited by John Krige and Dominique Pestre, 542–63. London: Routledge.

Kamminga, Harmke, and Andrew Cunningham, eds. 1995. *The Science and Culture of Nutrition, 1840–1940.* Amsterdam: Rodopi.

Kandel, Eric R. 1983. "Neurobiology and Molecular Biology: The Second Encounter." *Cold Spring Harbor Symposia on Quantitative Biology* 48: 891–908.

Kant, Immanuel. 1983 [1793]. *Kritik der Urteilskraft.* In *Kritik der Urteilskraft und Schriften zur Naturphilosophie,* vol. 5, edited by Wilhelm Weischedel. Darmstadt: Wissenschaftliche Buchgesellschaft.

Kaplan, Sarah, and Joanna Radin. 2011. "Bounding an Emerging Technology: Para-scientific Media and the Drexler-Smalley Debate about Nanotechnology." *Social Studies of Science* 41, no. 4: 457–85.

Kay, Lily E. 1993. *The Molecular Vision of Life: Caltech, The Rockefeller Foundation and the Rise of the New Biology.* Oxford: Oxford University Press.

Kay, Lily E. 2000. *Who Wrote the Book of Life? A History of the Genetic Code.* Stanford: Stanford University Press.

Keller, Evelyn Fox. 2002. *Making Sense of Life: Explaining Biological Development with Models, Metaphors, and Machines.* Cambridge, MA: Harvard University Press.

Keller, Evelyn Fox. 2009a. *Self-organization, Self-assembly, and the Inherent Activity of Matter.* The Hans Rausing Lecture 2009. Uppsala: Uppsala Universitet.

Keller, Evelyn Fox. 2009b. "Organisms, Machines, and Thunderstorms: A History of Self-Organization, Part Two. Complexity, Emergence, and Stable Attractors." *Historical Studies in the Natural Sciences* 39, no. 1: 1–31.

Keller, Evelyn Fox. 2011. "Towards a Science of Informed Matter." *Studies in History and Philosophy of Science Part C: Studies in History and Philosophy of Biological and Biomedical Sciences* 42, no. 2: 174–9.

Keller, Evelyn Fox. 2016. "Active Matter, Then and Now." *History and Philosophy of the Life Sciences* 38: 11.

Kennedy, Eugene P. 1966. "Biochemical Aspects of Membrane Function." In *Current Aspects of Biochemical Energetics: Fritz Lipmann Dedicatory Volume*, edited by Nathan O. Kaplan and Eugene P. Kennedy, 433–45. New York: Academic Press.

Khorana, H. Gobind. 1960. "Synthesis of Nucleotides, Nucleotide Coenzymes and Polynucleotides." *Federation Proceedings* 19: 931–40.

Khorana, H. Gobind. 1980. "Chemical Studies of Biological Membranes." *Bioorganic Chemistry* 9, no. 3: 363–405.

Khorana, H. Gobind. 1988. "Bacteriorhodopsin, a Membrane-Protein That Uses Light to Translocate Protons." *Journal of Biological Chemistry* 263, no. 16: 7439–42.

Khorana, H. Gobind, ed. 2000. *Chemical Biology: Selected Papers by Har Gobind Khorana with Introductions.* Singapore: World Scientific Publishing.

Kirschner, Marc, John Gerhart, and Tim Mitchison. 2000. "Molecular 'Vitalism.'" *Cell* 100: 79–88.

Klein, Ursula, and Wolfgang Lefèvre. 2007. *Materials in Eighteenth-Century Science. A Historical Ontology.* Cambridge, MA: MIT Press.

Klingenberg, Martin. 2005. "When a Common Problem Meets an Ingenious Mind: The Invention of the Modern Micropipette." *EMBO Reports* 6, no. 9: 797–800.

Klug, Aaron. 1983. "From Macromolecules to Biological Assemblies (Nobel Lecture)." *Angewandte Chemie International Edition* 22, no. 8: 565–82.

Kohler, Robert E. 1973. "The Enzyme Theory and the Origin of Biochemistry." *Isis* 64, no. 2: 181–96.

Koselleck, Reinhart. 1989. "'Erfahrungsraum' und 'Erwartungshorizont'—zwei historische Kategorien." In *Vergangene Zukunft. Zur Semantik geschichtlicher Zeiten*, 349–75. Frankfurt/M.: Suhrkamp.

Kornberg, Arthur. 1989. *For the Love of Enzymes: The Odyssey of a Biochemist.* Cambridge, MA: Harvard University Press.

Krebs, Hans, and Roswitha Schmid. 1979. *Otto Warburg: Zellphysiologe, Biochemiker, Mediziner.* Reihe Große Naturforscher no. 41. Stuttgart: Wissenschaftliche Verlagsgesellschaft.

Kremer, Richard L. 1997. "The Eye as Inscription Device in the 1870s: Optograms, Cameras, and the Photochemistry of Vision." In *Biology Integrating Scientific Fundamentals: Contributions to the History of Interrelations between Biology, Chemistry, and Physics from the 18th to the 20th Centuries*, edited by Brigitte Hoppe, 320–31. München: Institut für Geschichte der Naturwissenschaften.

Krohs, Ulrich. 2012. "Convenience Experimentation." *Studies in History and Philosophy of Science Part C: Studies in History and Philosophy of Biological and Biomedical Sciences* 43, no. 1: 52–7.

Krumhansl, James A., and Yoh-Ha Pao. 1979. "Microscience: An Overview." *Physics Today* 11: 25–32.

Kühlbrandt, Werner. 2000. "Bacteriorhodopsin—the Movie." *Nature* 406: 569–70.

Kühlbrandt, Werner. 2014. "The Resolution Revolution." *Science* 343: 1443.

Kuhn, Hans. 1965. "Versuche zur Herstellung einfacher organisierter Systeme von Molekülen." *Verhandlungen der Schweizerischen Naturforschenden Gesellschaft* 145: 245–66.

Kuhn, Hans. 1987. "Self-Organizing Molecular Electronic Devices." In *Molecular Electronic Devices II*, edited by Forrest L. Carter, 411–26. New York: Marcel Dekker.

Landecker, Hannah. 2013. "Metabolism, Reproduction, and the Aftermath of Categories." *Feminist & Scholar Online* 11, no. 3, online edition.

Lanyi, Janos K. 2004. "Bacteriorhodopsin." *Annual Review of Physiology* 66: 665–88.

Laszlo, Pierre. 2006. "Synthèse." In *Dictionnaire d'Histoire et Philosophie des Sciences*, edited by Dominique Lecourt, 1045–47. 4th ed. Paris: PUF.

Latour, Bruno, and Steve Woolgar. 1979. *Laboratory Life: The Construction of Scientific Facts*. Princeton: Princeton University Press.

Lécuyer, Christophe, and David C. Brock. 2006. "The Materiality of Microelectronics." *History and Technology* 22: 301–25.

Lefkowitz, Robert J. 2004. "Historical Review: A Brief History and Personal Retrospective of Seven-Transmembrane Receptors." *Trends in Pharmacological Sciences* 25, no. 8: 413–22.

Lehninger, Albert L. 1970. *Biochemistry: The Molecular Basis of Cell Structure and Function*. New York: Worth Publishers.

Leitenberg, Milton, and Raymond A. Zilinskas, with Jens H. Kuhn. 2012. *The Soviet Biological Weapons Program: A History*. Cambridge: Harvard University Press.

Levina, Elena S., and A. Sedov. 2000. "Molecular Biology in the Soviet Russia (An Essay)." *Molecular Biology* 34, no. 3: 420–47.

Lipan, Ovidiu, and Wing H. Wong. 2006. "Is the Future Biology Shakespearean or Newtonian?" *Molecular Biosystems* 2: 411–16.

Lipmann, Fritz. 1941. "Metabolic Generation and Utilization of Phosphate Bond Energy." *Advances in Enzymology and Related Areas of Molecular Biology* 1: 99–162.

Lipmann, Fritz. 1971. "Molecular Technology." In *Wanderings of a Biochemist*, 176–87. New York: Wiley.

Loeb, Jacques. 1915. "Mechanistic Science and Metaphysical Romance." *Yale Review* 4: 766–85.

Lozier, Richard H., Roberto A. Bogomolni, and Walther Stoeckenius. 1975. "Bacteriorhodopsin: A Light-Driven Proton Pump in *Halobacterium halobium*." *Biophysical Journal* 15: 955–62.

Liu, Daniel. 2018. "Heads and Tails: Molecular Imagination and the Lipid Bilayer, 1917–1941." In *Visions of Cell Biology: Reflections Inspired by Cowdry's General Cytology*, edited by K. S. Matlin, J. Maienschein, and M. D. Laubichler, 209–45. Chicago: University of Chicago Press.

Loeve, Sacha. 2015. "La Notion d'Objet Relationnel dans les Nanotechnologies (avec et après Simondon)." In *Cahiers Simondon*, no. 6, edited by Jean-Hugues Barthélémy, 47–109. Paris: L'Harmattan.

Loeve, Sacha. 2016. "L'ATP-synthase: Un Moteur Moléculaire? (Petit Récit Technique)." In *Gilbert Simondon ou l'Invention du Futur*, edited by V. Bontemps, 131–45. Paris: Klinksieck.

Lombard, Jonathan. 2014. "Once Upon a Time the Cell Membrane: 175 Years of Cell Boundary Research." *Biology Direct* 9, no. 1: 32.

Lynen, Feodor. 1964. "The Pathway from 'Activated Acetic Acid' to the Terpenes and Fatty Acids." *Proceedings of the Royal Caroline Institute* 103–38.

Maniatis, Tom, Edward F. Fritsch, and Joseph Sambrook. 1982. *Molecular Cloning: A Laboratory Manual*. Cold Spring Harbor: Cold Spring Harbor Laboratory.

Margulis, Lynn, and Dorion Sagan. 1986. *Microcosmos: Four Billion Years of Evolution from Our Microbial Ancestors*. New York: Summit Books.

Marschall, Luitgard. 2000. *Im Schatten der chemischen Synthese: Industrielle Biotechnologie in Deutschland (1900–1970)*. München: Campus.

Matsuno, Koichiro. 1991. "Molecular Electronics: Biosensors and Biocomputers. Edited by Felix T. Hong; Published by Plenum Press, New York, 1989, US$95.00" [Book Review]. *Biosystems* 24, no.4: 323–24.

Maunsbach, Arvid B. 2008. "A Profile of Fritiof S. Sjöstrand—The Founding Editor." *Journal of Structural Biology* 163, no. 3: 196–200.

Maynard-Smith, John, and Eörs Szathmáry. 1995. *The Major Transitions in Evolution*. New York: W. H. Freeman.

McAuliffe, Kathleen. 1981. "Biochip Revolution." *Omni*, December, 52–8.

McAuliffe, Kathleen. 1984. "Smart Cells." *Omni*, March, 119–20.

McCray, W. Patrick. 2013. *The Visioneers: How a Group of Elite Scientists Pursued Space Colonies, Nanotechnologies, and a Limitless Future*. Princeton: Princeton University Press.

Merrifield, Robert Bruce. 1985. "Solid Phase Synthesis (Nobel Lecture)." *Angewandte Chemie International Edition in English* 24, no. 10: 799–810.

Miyasaka, Tsutomu, Koichi Koyama, and Isamu Itoh. 1992. "Quantum Conversion and Image Detection by a Bacteriorhodopsin-Based Artificial Photoreceptor." *Science* 255, no. 5042: 342–44.

Mody, Cyrus C. M. 2011. *Instrumental Community: Probe Microscopy and the Path to Nanotechnology*. Cambridge, MA: MIT Press.

Mody, Cyrus C. M. 2016. "Professional Scientist." In *A Companion to the History of Science*, edited by Bernard Lightman, 164–77. Chichester: John Wiley & Sons.

Mody, Cyrus C. M. 2017a. *The Long Arm of Moore's Law: Microelectronics and American Science*. Cambridge, MA: MIT Press.

Mody, Cyrus C. M. 2017b. "Square Scientists and the Excluded Middle." *Centaurus* 59: 58–71.

Monod, Jacques. 1971. *Zufall und Notwendigkeit. Philosophische Fragen der modernen Biologie*. München: Piper.

Morange, Michel. 1998. *A History of Molecular Biology*. Cambridge, MA: Harvard University Press.

Morange, Michel. 2007. "What History Tells Us XI: The Complex History of the Chemiosmotic Theory." *Journal of Biosciences* 32: 1245–50.

Morange, Michel. 2008. "The Death of Molecular Biology?" *History and Philosophy of the Life Sciences* 30: 31–42.

Morange, Michel. 2009. "A Critical Perspective on Synthetic Biology." *HYLE: International Journal for the Philosophy of Chemistry* 15: 21–30.

Morange, Michel. 2011. "What History Tells Us XXV: Construction of the Ribbon Model of Proteins (1981). The Contribution of Jane Richardson." *Journal of Biosciences* 36, no. 4: 571–74.

Morange, Michel. 2013. "What History Tells Us XXX: The Emergence of the Fluid Mosaic Model of Membranes." *Journal of Biosciences* 38: 3–7.

Morgan, Neil. 1990. "The Strategy of Biological Research Programmes: Reassessing the 'Dark Age' of Biochemistry, 1910–1930." *Annals of Science* 47, no. 2: 139–50.

Morton, R. A., and G. A. J. Pitt. 1957. "Visual Pigments." *Fortschritte der Chemie organischer Naturstoffe* 14: 244–316.

Mueller, Paul, Donald O. Rudin, H. Ti Tien, and W. C. Wescott. 1962a. "Reconstitution of Cell Membrane Structure in Vitro and Its Transformation into an Excitable System." *Nature* 194: 979–80.

Mueller, Paul, Donald O. Rudin, H. Ti Tien, and W. C. Wescott. 1962b. "Reconstitution of Excitable Cell Membrane Structure in Vitro." *Circulation* 26, no. 5: 1167–71.

Müggenburg, Jan. 2014. "Clean by Nature: Lively Surfaces and the Holistic-Systemic Heritage of Contemporary Bionik." *communication +1*, no. 3, [article 9].

Müller-Wille, Staffan, and Hans-Jörg Rheinberger. 2012. *A Cultural History of Heredity*. Chicago: University of Chicago Press.

Myers, Natasha 2015. *Rendering Life Molecular: Models, Modelers, and Excitable Matter*. Durham: Duke University Press.

Nassal, M., T. Mogi, S. S. Karnik, and H. G. Khorana. 1987. "Structure-Function Studies on Bacteriorhodopsin, III: Total Synthesis of a Gene for Bacterio-opsin and Its Expression in *Escherichia Coli*." *Journal of Biological Chemistry* 262, no. 19: 9264–70.

Needham, Dorothy M. 1971. *Machina Carnis: The Biochemistry of Muscular Contraction in Its Historical Development*. Cambridge: Cambridge University Press.

Needham, Joseph. 1936. *Order and Life*. New Haven: Yale University Press.

Nicholson, Daniel J. 2013. "Organisms ≠ Machines." *Studies in History and Philosophy of Science Part C: Studies in History and Philosophy of Biological and Biomedical Sciences* 44, no. 4, Part B: 669–78.

Nickelsen, Kärin. 2015. *Explaining Photosynthesis: Models of Biochemical Mechanisms, 1840–1960*. Amsterdam: Springer Netherlands.

Nickelsen, Kärin, and Govindjee. 2011. *The Maximum Quantum Yield Controversy: Otto Warburg and the "Midwest-Gang."* Bern: Bern Studies in the History and Philosophy of Science.

November, Joseph A. 2012. *Biomedical Computing: Digitizing Life in the United States*. Baltimore: Johns Hopkins University Press.

Nunn, Jean. 1976. "Molecular Biology and Biochemistry. A Third Level Course. S322 Oxidative Phosphorylation: Programme 2." Milton Keynes: The Open University, VHS cassette, 27 min.

Nyhart, Lynn K. 2017. "The Political Organism. Carl Vogt on Animals and States in the 1840s and '50s." *Historical Studies in the Natural Sciences* 47, no. 5: 602–28.

Oesterhelt, Dieter. 1967. *Zur Kenntnis der Fettsäuresynthetase aus Hefe.* Dissertationsschrift. Ludwig-Maximilians-Universität München.

Oesterhelt, Dieter. 1972. "Die Purpurmembran aus *Halobacterium halobium.*" *Hoppe-Seylers Zeitschrift für physiologische Chemie* 353: 1554–55.

Oesterhelt, Dieter, Christoph Bräuchle, and Norbert Hampp. 1991. "Bacteriorhodopsin: A Biological Material for Information Processing." *Quarterly Reviews of Biophysics* 24, no. 4: 425–78.

Oesterhelt, Dieter, and Benno Hess. 1973. "Reversible Photolysis of the Purple Complex in the Purple Membrane of *Halobacterium halobium.*" *European Journal of Biochemistry* 37, no. 2: 316–26.

Oesterhelt, Dieter, Susanne Meeßen, Jörg Tittor, Anja Matuszat, and Klaus May. 1990. Bakteriorhodopsin-Doppelmutanten. German patent DE4037342A1, filed Nov. 23rd, 1990, and issued May 27th, 1992.

Oesterhelt, Dieter, Jörg Soppa, and Günter Krippahl. 1987. *Halobacterium*-Stämme, Bakteriorhodopsin-Modifikationen und Herstellungsverfahren. German patent DE000003730424A1, filed Sept. 10th, 1987, and issued Mar. 23rd, 1989.

Oesterhelt, Dieter, and Walther Stoeckenius. 1971. "Rhodopsin-like Protein from the Purple Membrane of *Halobacterium halobium.*" *Nature New Biology* 233, no. 39: 149–52.

Oesterhelt, Dieter, and Walther Stoeckenius. 1973. "Functions of a New Photoreceptor Membrane." *Proceedings of the National Academy of Sciences of the United States of America* 70, no. 10: 2853–57.

Olby, Robert. 1986. "Structural and Dynamic Explanations in the World of Neglected Dimensions." In *A History of Embryology: Eighth Symposium of the British Society for Developmental Biology,* edited by T. J. Horder, J. A. Witkowski, and C. C. Wylie, 175–203. Cambridge: Cambridge University Press.

Olby, Robert. 1990. "The Molecular Revolution in Biology." In *Companion to the History of Modern Science,* edited by Robert C. Olby, Geoffrey N. Cantor, John R. R. Christie, and M. Jonathan S. Hodge, 503–20. London: Routledge.

Oprian, Daniel D., Michael Nassal, Luca Ferretti, Sadashiva S. Karnik, and H. Gobind Khorana. 1986. "Design and Total Synthesis of a Gene for Bovine Rhodopsin." In *Protein Engineering: Applications in Science, Medicine, and Industry,* edited by M. Inouye, and R. Sarma, 111–23. New York: Academic Press.

Otis, Laura. 1999. *Membranes: Metaphors of Invasion in Nineteenth-Century Literature, Science, and Politics.* Baltimore: Johns Hopkins University Press.

Ovchinnikov, Y. A., N. G. Abdulaev, M. Y. Feigina, A. V. Kiselev, and N. A. Lobanov. 1979. "Structural Basis of the Functioning of Bacteriorhodopsin— Overview." FEBS *Letters* 100, no. 2: 219–24.

Pardee, Arthur B. 1968. "Membrane Transport Proteins." *Science* 162, no. 3854: 632–37.

Pothier, Richard. 1972. "Life's Energy System Is Partially Duplicated." *Miami Herald*, January 15, 1972.

Prebble, John N. 2013. "Contrasting Approaches to a Biological Problem: Paul Boyer, Peter Mitchell and the Mechanism of the ATP Synthase, 1961–1985." *Journal of the History of Biology* 46, no. 4: 699–737.

Prebble, John, and Bruce Weber. 2003. *Wandering in the Gardens of the Mind: Peter Mitchell and the Creation of Glynn*. Oxford: Oxford University Press.

Prigogine, Ilya, and Isabelle Stengers. 1984. *Order from Chaos*. London: Heinemann.

Projektleitung Biologie, Ökologie, Energie, ed. 1989. *Programm Angewandte Biologie und Biotechnologie. Jahresbericht 1986*. Jülich.

Prüll, Cay-Rüdiger, Andreas-H. Maehle, and Robert F. Halliwell. 2009. *A Short History of the Drug Receptor Concept*. Basingstoke: Palgrave Macmillan.

Pullman, Bernard, and Alberte Pullman. 1963. *Quantum Biochemistry*. New York: Wiley Interscience.

Rabinbach, Anson. 1992. *The Human Motor: Energy, Fatigue, and the Origins of Modernity*. Berkeley: University of California Press.

Rabinow, Paul. 2006. *Making PCR: A Story of Biotechnology*. Chicago: University of Chicago Press.

Racker, Efraim. 1965. *Mechanisms in Bioenergetics*. New York: Academic Press.

Racker, Efraim. 1974. "Mechanism of ATP Formation in Mitochondria and Ion Pumps." In *Dynamics of Energy-Transducing Membranes*, edited by Lars Ernster, R. W. Estabrook, E. C. Slater, 269–81. Amsterdam: Elsevier.

Racker, Efraim. 1976. *A New Look at Mechanisms in Bioenergetics*. New York: Academic Press.

Racker, Efraim. 1977. "Mechanisms of Energy Transformations." *Annual Review of Biochemistry* 46, no.1: 1006–14.

Racker, Efraim, and Franziska W. Racker. 1981. "Resolution and Reconstitution: A Dual Autobiographical Sketch." In *Evolving Life Sciences—Of Oxygen, Fuels, and Living Matter*, vol. 1, edited by Giorgio Semenza, 265–87. Chichester: John Wiley & Sons.

Racker, Efraim, and Walther Stoeckenius. 1974. "Reconstitution of Purple Membrane Vesicles Catalyzing Light-Driven Proton Uptake and Adenosine Triphosphate Formation." *Journal of Biological Chemistry* 249: 662–63.

Ramaswamy, Sriram. 2010. "The Mechanics and Statistics of Active Matter." *Annual Review of Condensed Matter Physics* 1: 323–45.

Rasmussen, Nicolas. 1997. *Picture Control: The Electron Microscope and the Transformation of Biology in America, 1940–1960*. Stanford: Stanford University Press.

Rasmussen, Nicolas. 2014. *Gene Jockeys: Life Science and the Rise of Biotech Enterprise*. Baltimore: Johns Hopkins University Press.

Rebentrost, Inken. 2006. *Das Labor in der Box: Technikentwicklung und Unternehmensgründung in der frühen deutschen Biotechnologie*. München: C. H. Beck.

Reichstein, Tadeusz. 1985. Oral history interview by Tonja Koeppel, Basel, Switzerland, April 22, 1985, transcript #0040. Philadelphia: Chemical Heritage Foundation.

Reinhardt, Carsten. 2006. *Shifting and Rearranging: Physical Methods and the Transformation of Modern Chemistry.* Sagamore Beach: Science History Publications/USA.

Reinhardt, Carsten. 2014. "The Olfactory Object: Towards a History of Smell in the 20th Century." In *Objects of Chemical Inquiry,* edited by Ursula Klein and Carsten Reinhardt, 321–41. Sagamore Beach: Science History Publications/USA.

Reinhardt, Carsten. 2017. "'This Other Method': The Dynamics of NMR in Biochemistry and Molecular Biology." *Historical Studies in the Natural Sciences* 47, no. 3: 389–422.

Reynolds, Andrew S. 2007. "The Cell's Journey: From Metaphorical to Literal Factory." *Endeavour* 31: 65–70.

Reynolds, Andrew S. 2018. *The Third Lens: Metaphor and the Creation of Modern Cell Biology.* Chicago: University of Chicago Press.

Rheinberger, Hans-Jörg. 1995. "Beyond Nature and Culture: A Note on Medicine in the Age of Molecular Biology." *Science in Context* 8, no. 1: 249–63.

Rheinberger, Hans-Jörg. 1997. *Toward a History of Epistemic Things: Synthesizing Proteins in the Test Tube.* Stanford: Stanford University Press.

Rheinberger, Hans-Jörg. 2015. "Preparations, Models, and Simulations." *History and Philosophy of the Life Sciences* 36, no. 3: 321–34.

Riskin, Jessica. 2016. *The Restless Clock: A History of the Centuries-Long Argument over What Makes Living Things Tick.* Chicago: University of Chicago Press.

Robinson, Joseph D. 1997. *Moving Questions. A History of Membrane Transport and Bioenergetics.* Oxford, New York: Oxford University Press.

Roland, Alex, and Philip Shiman. 2002. *Strategic Computing: DARPA and the Quest for Machine Intelligence, 1983–1993.* Cambridge, MA: MIT Press.

Roosth, Sophia. 2017. *Synthetic: How Life Got Made.* Chicago: University of Chicago Press.

Rose, Nikolas. 2009. *The Politics of Life Itself: Biomedicine, Power, and Subjectivity in the Twenty-First Century.* Princeton: Princeton University Press.

"Roter Bazillus." 1992. *Der Spiegel* 7: 213–14.

The Royal Swedish Academy of Sciences. 2017. "*Scientific Background on the Nobel Prize in Chemistry 2017. The Development of Cryo-electron Microscopy.*" Oct. 4th, 2017. https://www.nobelprize.org/nobel_prizes/chemistry/laureates/2017/advanced-chemistryprize2017.pdf.

Russell, C.A. 1987. "The Changing Role of Synthesis in Organic Chemistry." *Ambix* 34, no. 3: 169–80.

Sapp, Jan. 2015. "On Symbiosis, Microbes, Kingdoms and Domains." In *Earth, Life and System: Evolution and Ecology on a Gaian Planet,* edited by B. Clarke, 105–26. New York: Fordham University Press.

Schäffner, Wolfgang. 2017. "Active Matter." In *23 Manifeste zu Bildakt und Verkörperung,* edited by Pablo Schneider and Marion Lauschke, 1–10. Berlin: De Gruyter.

Schatz, Gottfried. 1996. "Efraim Racker." *Biographical Memoirs* 70: 321–49.

Schick, G. Alan, Albert F. Lawrence, and Robert R. Birge. 1988. "Biotechnology and Molecular Computing." *Trends in Biotechnology* 6, no. 7: 159–63.

Schlögl, Reinhard. 1967. "Membranen als Schranken, Schleusen und Pumpen für den molekularen Transport." *Berichte der Bunsengesellschaft für physikalische Chemie* 71, no. 8 (October): 755–58.

Service, Robert F., and Erik Stokstad. 2017. "Cold, Clear View of Molecules Nets Chemistry Prize." *Science* 358, no. 6360: 156.

Shapin, Steven. 1989. "The Invisible Technician." *American Scientist* 77, no. 6: 554–63.

Shapin, Steven. 2008. *The Scientific Life: A Moral History of a Late Modern Vocation.* Chicago: University of Chicago Press.

Shapiro, T. Rees. 2011. "Biological Weapons Expert Tucker, 56, Was Known for Fluency in Politics." *Washington Post*, August 3, 2011.

Shuman, Howard A. 2003. "Just Toothpicks and Logic: How Some Labs Succeed at Solving Complex Problems." *Journal of Bacteriology* 185, no. 2: 387–90.

Sinding, Christiane. 2006. "Récepteur." In *Dictionnaire d'Histoire et Philosophie des Sciences*, edited by Dominique Lecourt, 491–93. 4th ed. Paris: PUF.

Singer, S.J., and Garth L. Nicolson. 1972. "The Fluid Mosaic Model of the Structure of Cell Membranes." *Science* 175, no. 4023: 720–31.

Sjöstrand, Fritiof S. 1963. "A Comparison of Plasma Membrane, Cytomembranes, and Mitochondrial Membrane Elements with Respect to Ultrastructural Features." *Journal of Ultrastructure Research* 9, no. 5–6: 561–80.

Slater, Leo B. 2002. "Instruments and rules: R. B. Woodward and the tools of twentieth-century organic chemistry." *Studies in History and Philosophy of Science Part A* 33, no. 1: 1–33.

Slater, Leo B. 2008. "Robert Burns Woodward (1917–1979)." In *New Dictionary of Scientific Biography*, edited by Noretta Koertge, 349–56. Detroit: Charles Scribner's Sons/Thomson Gale.

Soentgen, Jens. 2008. "Stuff: A Phenomenological Definition." In *Stuff: The Nature of Chemical Substances*, edited by Klaus Ruthenberg and Jaap van Brakel, 71–91. Würzburg: Königshausen & Neumann.

Span, Paula. 2016. "Study Finds Growing Reason to Be Wary of Some Reflux Drugs." *New York Times*, Jan. 29, 2016.

Stadler, Max. 2010. "Assembling Life: Models, the Cell, and the Reformations of Biological Science, 1920–1960." PhD diss., University of London.

Stadler, Max. 2014. "Neurohistory Is Bunk? The Not-So-Deep History of the Postclassical Mind." *Isis* 105, no. 1: 133–44.

Steinbach, H. Burr. 1940. "Electrolyte Balance of Animals Cells." *Cold Spring Harbor Symposia on Quantitative Biology* 8: 242–54.

Steinhauser, Thomas. 2014. *Zukunftsmaschinen in der Chemie: Kernmagnetische Resonanz bis 1980.* Frankfurt a.M.: Peter Lang.

Steinhauser, Thomas, Jeremiah James, Dieter Hoffmann, and Bretislav Friedrich. 2011. *Hundert Jahre an der Schnittstelle von Chemie und Physik: Das Fritz-Haber-Institut der Max-Planck-Gesellschaft zwischen 1911 und 2011.* Berlin: De Gruyter.

Stevens, Hallam. 2013. *Life out of Sequence: A Data-driven History of Bioinformatics.* Chicago: University of Chicago Press.

Stiftung Werner-von-Siemens-Ring, ed. 2001. "Werner-von-Siemens-Ring. Verleihung an Dieter Oesterhelt 1999." *Schriften der Stiftung Werner-von-Siemens-Ring* 19.

Stoeckenius, Walther. 1971. "Structure and Composition of the Purple Membrane, a Differentiated Region of the Cell Membrane of *Halobacterium halobium*." *Federation Proceedings* 30, no. 3: 1031.

Stoeckenius, Walther. 1976. "The Purple Membrane of Salt-Loving Bacteria." *Scientific American*, June, 38–47.

Stoeckenius, Walther. 1994. "From Membrane Structure to Bacteriorhodopsin." *Journal of Membrane Biology* 139: 139–48.

Stoeckenius, Walther, and Wolf H. Kunau. 1968. "Further Characterization of Particulate Fractions from Lysed Cell Envelopes of *Halobacterium halobium* and Isolation of Gas Vacuole Membranes." *Journal of Cell Biology* 38, no. 2: 337–57.

Stoeckenius, Walther, and Richard H. Lozier. 1974. "Light Energy Conversion in *Halobacterium halobium*." *Journal of Supramolecular Structure* 2 no. 5–6: 769–74.

Stoeckenius, Walther, Richard H. Lozier, and Roberto A. Bogomolni. 1979. "Bacteriorhodopsin and the Purple Membrane of Halobacteria." *Biochimica et Biophysica Acta (BBA)—Reviews on Bioenergetics* 505, no. 3–4: 215–78.

Stoeckenius, Walther, and Robert Rowen. 1967. "A Morphological Study of *Halobacterium halobium* and Its Lysis in Media of Low Salt Concentration." *Journal of Cell Biology* 34, no. 1: 365–91.

Stoff, Heiko. 2012. *Wirkstoffe: Eine Wissenschaftsgeschichte der Hormone, Vitamine und Enzyme, 1920–1970*. Stuttgart: Franz Steiner Verlag.

Strasser, Bruno J. 2012. "Collecting Nature: Practices, Styles, and Narratives." *Osiris* 27, no. 1: 303–40.

Strasser, Bruno J., and Soraya de Chadarevian. 2011. "The Comparative and the Exemplary: Revisiting the Early History of Molecular Biology." *History of Science* 49: 317–36.

Strick, James E. 2012. "A History of Origin of Life Ideas from Darwin to NASA." In *Genesis—In The Beginning: Precursors of Life, Chemical Models and Early Biological Evolution*, edited by J. Seckbach, 907–21. Dordrecht: Springer Netherlands.

Strick, James E. 2015. "Exobiology at NASA: Incubator for the Gaia and Serial Endosymbiosis Theories". In *Earth, Life and System: Evolution and Ecology on a Gaian Planet*, edited by B. Clarke, 80–104. New York: Fordham University Press.

Sullivan, Walter. 1976. "Bacteria Viewed as Power Source." *New York Times*, Mar. 3, 1976.

Sumper, Manfred, Heribert Reitmeier, and Dieter Oesterhelt. 1976. "Biosynthesis of the Purple Membrane of Halobacteria." *Angewandte Chemie International Edition in English* 15: 187–94.

"Synthetic Gene Works Well in Living Cell." 1976. *New Scientist*, Sept. 2, 475.

Szent-Györgyi, Albert. 1957. *Bioenergetics*. New York: Academic Press.

Tanford, Charles, and Jaqueline Reynolds. 2001. *Nature's Robots: A History of Proteins*. Oxford: Oxford University Press.

Teich, Mikulas. 1973. "From 'Enchyme' to 'Cyto-Skeleton': The Development of Ideas on the Chemical Organization of Living Matter." In *Changing Perspectives in the History of Science: Essays in Honour of Joseph Needham*, edited by Mikuláš Teich and Robert Maxwell Young, 439–71. London: Heinemann.

Teorell, Torsten. 1967. "Problems and Perspectives in Membranology." *Berichte der Bunsengesellschaft für physikalische Chemie* 71, no. 8: 814–17.

Tibbits, Skylar. 2017. "An Introduction to Active Matter." In *Active Matter*, edited by Skylar Tibbits, 11–18. Cambridge, MA: MIT Press.

Toepfer, Georg. 2011. "Organismus." In *Historisches Wörterbuch der Biologie*, vol. 2, 777–842. Stuttgart: Metzler.

Tracey, Kevin. 2015. "Shock Medicine: Stimulation of the Nervous System Could Replace Drugs for Inflammatory and Autoimmune Conditions." *Scientific American*, March, 30–35.

Trumpler, Maria. 1997. "Converging Images: Techniques of Intervention and Forms of Representation of Sodium-Channel Proteins in Nerve Cell Membranes." *Journal of the History of Biology* 30: 55–89.

Tucker, Jonathan B. 1984. "Biochips: Can Molecules Compute?" *High Technology* 2: 36–47.

Turner, Fred. 2006. *From Counterculture to Cyberculture: Stewart Brand, the Whole Earth Network and the Rise of Digital Utopianism*. Chicago: University of Chicago Press.

Ulmer, Kevin. 1982. "Biological Assembly of Molecular Ultracircuits." In *Molecular Electronic Devices*, edited by Forrest L. Carter, 213–22. New York: Marcel Dekker.

Unwin, P. Nigel T. 1971. "Phase Contrast and Interference Microscopy with the Electron Microscope." *Philosophical Transactions of the Royal Society B: Biological Sciences* 261: 95–104.

Unwin, P. Nigel T., and Richard Henderson. 1975. "Molecular Structure Determination by Electron Microscopy of Unstained Crystalline Specimens." *Journal of Molecular Biology* 94, no. 3: 425–32.

Vettel, Eric J. 2006. *Biotech: The Countercultural Origins of an Industry*. Philadelphia: University of Pennsylvania Press.

Vinothkumar, Kutti R., and Richard Henderson. 2010. "Structures of Membrane Proteins." *Quarterly Reviews of Biophysics* 43, no. 1: 65–158.

Vsevolodov, Nikolai N. 1998. *Biomolecular Electronics: An Introduction via Photosensitive Proteins*. Boston: Birkhäuser.

Vsevolodov, Nikolai N., G. N. Ivanitsky, M. S. Soskin, and V. B. Taranenko. 1986. "Biochrome Film Is a Reversible Medium for Optical Recording." [In Russian.] *Avtometriya* 2: 43–48.

Wacker Chemie, ed. 1991. "Die Wacker-Chemie GmbH berichtet über das Geschäftsjahr 1990. Geschäftsbericht 1990." Business report. München: Wacker Chemie GmbH.

Wacker Chemie, ed. 1992. "Die Wacker-Chemie GmbH berichtet über das Geschäftsjahr 1991. Geschäftsbericht 1991." Business report. München: Wacker Chemie GmbH.

Wacker Unternehmenskommunikation, ed. 2003. "100 Jahre Wacker Forschung—es gibt noch so viel zu erfinden." Company communication. München: Wacker Chemie GmbH, c.2003.

Wald, George. 1970. "Vision and the Mansion of Life: The First Feodor Lynen Lecture." In *Miami Winter Symposia*, 1–32. Amsterdam: North Holland Publishing.

Walz, Dieter, Shimon Schuldiner, and Mordhay Avron. 1971. "Photoreactions of Chloroplasts in a Glycine Medium." *European Journal of Biochemistry* 22, no. 3: 439–44.

Watsuji, Toru, Kazuro Nishi, Tetsuo Moriya, and Shuku Maeda. 1995. "Introduction to the Bioelectronic Devices Project in Japan." *BioSystems* 35, no. 2–3: 101–6.

Weber, Marcel. 2002. "Theory Testing in Experimental Biology: The Chemiosmotic Mechanism of ATP Synthesis." *Studies in History and Philosophy of Science Part C: Studies in History and Philosophy of Biological and Biomedical Sciences* 33, no. 1 (March): 29–52.

Weber, Marcel. 2004. *Philosophy of Experimental Biology*. Cambridge: Cambridge University Press.

Weiner, M. 1993. "Revolution in der Materialforschung, Philipp Morris Forschungspreis." Brochure of the Philipp Morris award, private collection.

White, David. 2000. *The Physiology and Biochemistry of Prokaryotes*. 2nd ed. Oxford: Oxford University Press.

Widdus, Roy, and Charles R. Ault. 1974. "Progress in Research Related to Genetic Engineering and Life Synthesis." *International Review of Cytology* 38: 7–66.

Wilkins, Maurice. 2003. *The Third Man of the Double Helix: The Autobiography of Maurice Wilkins*. Oxford: Oxford University Press.

Will, Heike. 2011. *Sei Naiv Und Mach' Ein Experiment. Feodor Lynen. Biographie des Münchner Biochemikers und Nobelpreisträgers*. Weinheim: Wiley-VCH.

Winnacker, Ernst-Ludwig. 1990. "Synthetische Biologie." In *Die zweiter Schöpfung. Geist und Ungeist in der Biologie des 20. Jahrhunderts*, edited by R. Hohlfeld and J. Herbig, 369–85. Hanser: München.

Wolstenholme, Gordon Ethelbert Ward, and Maeve O'Connor, eds. 1966. *Ciba Foundation Symposium: Principles of Biomolecular Organization*. London: Churchill.

Yanchinski, Stephanie. 1982. "And Now—The Biochip." *New Scientist* 93: 68–71.

Yi, Doogab. 2010. "The Integrated Circuit for Bioinformatics: The DNA Chip and Materials Innovation at Affymetrix." *Studies in Materials Innovation*. Philadelphia: Chemical Heritage Foundation.

Yi, D. 2015. *The Recombinant University: Genetic Engineering and the Emergence of Stanford Biotechnology*. Chicago: University of Chicago Press.

Zell, Rolf Andreas. 1994. "Gelb statt lila. Wie es zu einer Sternstunde der Forschung kommt." *Bild der Wissenschaft* 1: 22–27.

Index